Norman Lockyer

Elements of Astronomy

Norman Lockyer

Elements of Astronomy

ISBN/EAN: 9783337412234

Printed in Europe, USA, Canada, Australia, Japan

Cover: Foto ©berggeist007 / pixelio.de

More available books at **www.hansebooks.com**

ELEMENTS

OF

ASTRONOMY:

ACCOMPANIED WITH NUMEROUS ILLUSTRATIONS,
A COLORED REPRESENTATION OF THE SOLAR, STELLAR,
AND NEBULAR SPECTRA,

AND

CELESTIAL CHARTS OF THE NORTHERN AND THE
SOUTHERN HEMISPHERE.

BY

J. NORMAN LOCKYER,

FELLOW OF THE ROYAL ASTRONOMICAL SOCIETY, EDITOR OF "NATURE," ETC.

AMERICAN EDITION,

REVISED AND SPECIALLY ADAPTED TO THE SCHOOLS OF THE UNITED STATES.

NEW YORK ·:· CINCINNATI ·:· CHICAGO
AMERICAN BOOK COMPANY.
FROM THE PRESS OF
D. APPLETON & COMPANY.

ENTERED, according to Act of Congress, in the year 1870, by D. APPLETON & Co., in the Office of the Librarian of Congress, at Washington.

PREFACE.

THESE Elementary Lessons in Astronomy are intended, in the main, to serve as a text-book for use in Schools, but I believe they will be found useful to "children of a larger growth," who wish to make themselves acquainted with the basis and teachings of one of the most fascinating of the Sciences.

The arrangement adopted is new; but it is the result of much thought. I have been especially anxious in the descriptive portion to show the Sun's real place in the Cosmos, and to separate the real from the apparent movements. I have therefore begun with the Stars, and have dealt with the apparent movements in a separate chapter.

It may be urged that this treatment is objectionable, as it reduces the mental gymnastic to a minimum; it is right, therefore, that I should state that my aim throughout the book has been to give a connected view of the whole subject rather than to discuss any particular parts of it; and to supply facts, and ideas founded on the facts, to serve as a basis for subsequent study and discussion.

It has been my especial endeavor to incorporate the most recent astronomical discoveries. Spectrum-analysis and its results are therefore fully dealt with; and distances, masses, etc., are based upon the recent determination of the solar parallax.

The use of the Globes and that of the Telescope have both been touched upon. Now that our best opticians are employed in producing "Educational Telescopes," more than powerful enough for school purposes, at a low price, it is to be hoped that this aid to knowledge will soon find its place in every school, side by side with the blackboard and much questioning.

I take this opportunity of expressing my thanks to Mr. Warren De La Rue, and also to the Council of the Royal Astronomical Society, M. Guillemin, Mr. R. Bentley, the Rev. H. Godfray, Mr. Cooke, and Mr. Browning, who have kindly supplied me with many of the illustrations.

I am also under obligations to other friends, especially to Mr. Balfour Stewart and Mr. J. M. Wilson, for valuable advice and criticism, while the work has been passing through the press.

J. N. L.

It has been the aim of the American Editor to extend the usefulness of Mr. Lockyer's admirable treatise, by specially adapting it to the schools of the United States. With this view, he has condensed the text in some parts, and enlarged it in others; he has introduced new illustrations, and has added questions to facilitate the labors of the examiner and to furnish the student with a test of his preparation. He has also extended the practical directions for finding interesting objects in the heavens on different evenings throughout the year. Celestial Charts containing the constellations and principal stars, taken from the "Popular Astronomy" of Arago, have been appended. They will be found to answer the purpose of a large Atlas of the Heavens, and will help to inspire the learner with interest in the subject and to give a practical bearing to his studies.

It is hoped that this treatise, embodying as it does the most recent and interesting results of astronomical discovery—many of which are due to the researches of its distinguished author—will be found just what is needed as an elementary text-book on the subject, and that it may give an impetus to the study of Astronomy in both public and private schools throughout the land.

NEW YORK, *August*, 1870.

CONTENTS.

INTRODUCTION.

General View.—Usefulness of Astronomy.—Early History of Astronomy.—Mathematical Definitions, ... pp. 9-23

CHAPTER I.
THE STARS.

Magnitudes of the Stars.—Their Comparative Brightness, Distances, and Diameters.—The Milky Way.—The Magellanic Clouds.—Distribution of the Stars.—Division of the Stars into Constellations.—The Zodiacal, Northern, and Southern Constellations.—Names of the Stars.—The First-magnitude Stars.—Proper and Apparent Motion of the Stars, . . . pp. 23-36
Double, Triple and Quadruple Stars.—Multiple Stars.—Binary Stars.—Variable Stars.—Mira.—Algol.—Temporary Stars.—Cause of Variations of Brightness.—Colored Stars.—Changes of Color.—Structure of the Stars.—Materials of the Photospheres.—Causes of Color in the Stars.—Star Groups and Clusters, ... pp. 36-48

CHAPTER II.
NEBULÆ.

Nebulæ under the Telescope.—Irregular Nebulæ.—Ring and Elliptical Nebulæ.—Spiral Nebulæ.—Planetary Nebulæ.—Nebulæ surrounding Stars.—Brightness of the Nebulæ.—Variable Nebulæ.—Distribution of the Nebulæ.—Physical Constitution of the Nebulæ.—The Nebular Hypothesis, . pp. 48-54

CHAPTER III.
THE SUN.

The Sun's Disk.—Its Distance and Diameter, Volume and Mass.—Rotation of the Sun.—The Plane of the Ecliptic.—Inclination of the Sun's Axis.—Time of Rotation.—Telescopic Appearance of the Sun.—Sun-spots.—Faculæ.—Corrugations.—Willow-leaves.—Punctulations.—Red-flames and Prominences.—Explanation of the Appearances on the Disk.—The Sun, a Variable Star.—Elements in the Sun.—Benign Influences of the Sun.—Future of the Sun, ... pp. 55-70

CHAPTER IV.
THE SOLAR SYSTEM.

General Description.—Explanation of the Signs of the Planets.—Historical Details.—Discovery of Neptune.—Discovery of the Asteroids.—The Suspected

Planet, Vulcan.—Motions and Orbits of the Planets.—The Satellites.—Distances of the Planets from the Sun.—Comparative Size of the Planets.—Distances and Revolutions of the Satellites.—Volumes, Masses, and Densities of the Planets, ... pp. 70–81

CHAPTER V.

THE EARTH.

Shape of the Earth.—The Sensible Horizon.—Poles and Equator.—Proofs of the Earth's Rotation.—Foucault's Experiment.—The Gyroscope.—Imaginary Lines on the Earth's Surface.—Latitude and Longitude.—Zones.—Polar and Equatorial Diameter.—Motions of the Earth.—Velocity of the Earth's Motions.—Inclination of the Earth's Axis.—Succession of Day and Night.—Length of the Longest Day in Different Latitudes.—The Change of Seasons.—Difference of Time and Longitude.—How to determine Longitude at Sea, ... pp. 81–101

Structure of the Earth.—The Earth's Crust.—Stratified and Igneous Rocks.—The Interior of the Earth.—Density of the Earth's Crust.—The Flattening at the Poles explained.—The Earth's Atmosphere.—Winds, how produced.—Belts of Calms and Winds.—Clouds, Rain, Snow, Hail.—Chemical Elements of the Earth.—Composition of the Air.—Original Condition of the Earth, ... pp. 101–112

CHAPTER VI.

THE MOON.

Size of the Moon.—Its Distance from the Earth.—Its Period of Revolution.—Librations.—Nodes.—The Moon's Orbit.—Earth-shine.—The Moon's Light.—Telescopic Appearance of the Moon.—Lunar Craters.—The Crater Copernicus.—Walled Planes and Rifles.—Absence of Water and Atmosphere.—Rotation of the Moon.—Phases of the Moon, . . . pp. 113–124

CHAPTER VII.

ECLIPSES.

Explanation of Eclipses.—Total and Partial Eclipses of the Moon.—Total, Annular, and Partial Eclipses of the Sun.—Extent of a Partial Eclipse, how measured.—Recurrence of Eclipses.—The Saros.—Phenomena attending a Total Eclipse of the Sun.—The Corona.—Baily's Beads.—Luminous Prominences.—Number of Eclipses.—Memorable Eclipses.—Effects of Eclipses on the Uneducated, ... pp. 124–135

CHAPTER VIII.

THE INFERIOR AND SUPERIOR PLANETS.

The Planets distinguished as Inferior and Superior.—Mercury.—Its Phases, Orbit, and Apparent Diameter.—Its Heat and Light.—Its Rotation, Density, etc.—Venus.—Its Size and Density.—Its Seasons.—Its Day and Year.—Its Heat and Light, ... pp. 135–139

The Superior Planets.—Mars.—Its Day and Year.—Inclination of its Axis.—Its Apparent Diameter.—Its Appearance in 1862.—Its Atmosphere.—Its Seasons.—Jupiter, its Revolution, Rotation, Seasons, etc.—Its Belts.—Its Rotary Velocity.—Probability of a Great Cloudy Atmosphere.—Jupiter's Four Moons.—Saturn.—Its Size, Day, and Year.—Its Eight Moons.—Its Rings.—Its Seasons.—Uranus.—Its Size, Heat, and Light.—Its Four Moons.—Neptune and its Moon, ... pp. 139–152

CONTENTS.

CHAPTER IX.
THE ASTEROIDS, OR MINOR PLANETS.

Bode's Law.—Discovery of the Asteroids.—Size of the Asteroids.—Their Orbits.—Evidences of Atmosphere and Rotation.—Mode of detecting the Asteroids.—Theory respecting the Asteroids, . . . pp. 152-155

CHAPTER X.
COMETS.

General Description of Comets.—The Cometary Orbits.—Short-period Comets.—Long-period Comets.—Distances of Comets from the Sun.—Appearances presented by a Comet.—Changes in Appearance, as they approach the Sun.—Danger from Collision with a Comet.—A Divided Comet.—Physical Constitution of Comets.—Number of Comets.—Comets, how formerly regarded, ... pp. 155-164

CHAPTER XI.
METEORS AND METEORITES.

Number of Meteors.—The Zodiacal Light; its Shape, and Theory as to its Cause.—The November Meteoric Showers.—Orbits of the Meteors.—Cause of the Luminous Appearance of Meteors.—Their Size and Distance from the Earth.—Meteoric Showers of August and April.—Detonating Meteors.—Meteorites.—Showers of Aërolites.—Composition and Structure of Meteorites, ... pp. 164-173

CHAPTER XII.
APPARENT MOVEMENTS OF THE HEAVENLY BODIES.

The Earth, a Moving Observatory.—Apparent Movements, how produced.—The Celestial Sphere.—Celestial Poles and Equator.—Zenith and Nadir.—Declination and Right Ascension.—North-polar Distance.—The Horizon.—Altitude and Azimuth, ... pp. 173-179
Apparent Movements of the Stars.—Stars Visible in Different Latitudes.—Use of the Globes.—The Circumpolar Constellations.—Period of the Apparent Movements of the Celestial Sphere.—The Apparent Movements of the Stars, as affected by the Earth's Yearly Revolution.—How to identify the Stars in the Sky.—Constellations Visible in the United States on Different Evenings throughout the Year, ... pp. 179-194
Apparent Movements of the Sun.—Difference between the Sidereal and the Solar Day.—Celestial Latitude and Longitude.—Signs of the Zodiac.—Apparent Path of the Sun.—How to determine the Time of Sunrise and Sunset with the Celestial Globe.—How to find the Length of Day and Night.—Apparent Movements of the Moon.—The Harvest Moon, . . . pp. 194-202
Apparent Movements of the Planets.—Distances of the Planets from the Earth.—Phases of the Planets.—Aspects of the Planets; Conjunctions, Opposition, and Quadrature.—Transits.—Elongations.—Retrograde Motion, explained.—Synodic Period.—Inclinations of the Orbits, and Nodes.—Path of Venus among the Stars.—Representation of the Orbits of Mars and the Earth.—Saturn's Rings, as seen at Different Times from the Earth, . pp. 202-213

CHAPTER XIII.
THE MEASUREMENT OF TIME.

Clepsydræ.—Sun-dials.—Clocks and Watches.—The Mean Sun.—Irregularities of the Sun's Apparent Daily Motion.—Equation of Time.—The Apparent

Solar Day, Mean Solar Day, and Civil Day.—Sidereal Time.—The Week; Names of the Days.—The Month, Lunar, Tropical, Sidereal, Anomalistic, Nodical, and Calendar.—The Year, Sidereal, Solar, and Anomalistic.—The Calendar.—Old and New Style.—Change in the Length of the Solar Year.—Change of Aphelion and Perihelion, ... pp. 214-226.

CHAPTER XIV.
ASTRONOMICAL INSTRUMENTS.

Light.—Its Velocity.—Aberration of Light.—Its Reflection and Refraction.—Effect of Refraction.—Dispersion of Light.—The Spectrum.—Lenses.—Refraction by Convex and Concave Lenses.—Achromatic Lenses, pp. 226-235.

The Telescope.—Its Invention.—Its Construction.—Its Illuminating and Magnifying Power.—Eye-pieces.—The Largest Refractor.—Lord Rosse's Reflector.—Equatorial Telescopes.—Measurement of Angles.—The Altazimuth.—The Transit-circle.—Methods of determining the Time of Transit over a Wire.—Determination of Positions with the Equatorial.—Star-catalogues.—Corrections to be applied to Observations.—Parallax.—Changes in Positions already determined.—Precession of the Equinoxes.—Secular Variation of the Obliquity of the Ecliptic.—Celestial Latitude and Longitude, how determined, ... pp. 235-255.

Determination of Time, Latitude, and Longitude.—Determination of Distances.—The Moon's Parallax.—Determination of the Distance of Mars.—The Sun's Parallax, Old and New Value.—Parallax of the Stars.—Bessel's Method of determining the Distances of the Stars.—Table of the Parallax and Distances of some of the Nearest Stars.—Mode of determining the Size of the Heavenly Bodies, ... pp. 255-263.

CHAPTER XV.
THE SPECTRUM.

Gradual Formation of a Spectrum.—Fraunhofer's Lines.—Experiments with the Spectroscope.—Kirchhoff's Discovery.—Explanation of Fraunhofer's Lines.—Spectra of the Stars, Nebulæ, Moon, and Planets.—Explanation of the Frontispiece.—The Star Spectroscope.—Celestial Photography, . pp. 263-271.

CHAPTER XVI.
UNIVERSAL GRAVITATION.

Motion.—The Parallelogram of Forces.—Weight.—Laws of Falling Bodies.—Curvilinear Motion, how produced.—Newton's Discovery.—The Law of Gravity.—Effect of Gravity on the Moon's Path.—Kepler's Laws.—Centrifugal and Centripetal Force.—The Planetary Orbits.—Varying Velocity of a Body moving in an Elliptical Orbit, explained.—Gravity not Dependent on the Mass of the Attracted Body.—The Centre of Gravity, . pp. 272-283.

Determination of the Earth's Density and Mass.—The Cavendish Experiment.—Determination of the Sun's Mass.—Determination of the Masses of the Planets.—Perturbations and Inequalities.—Precession of the Equinoxes, how produced.—Change in the Earth's Axis.—Nutation.—Tides, how produced.—Velocity and Height of the Tidal Wave.—Effect of Tidal Action on the Daily Rotation, ... pp. 283-293.

APPENDIX.—Tables, ... pp. 294-299.
ALPHABETICAL INDEX, ... pp. 300-312.

ELEMENTS OF ASTRONOMY.

INTRODUCTION.

General View.

1. **Astronomy** is the science that treats of the heavenly bodies.

2. **The Heavenly Bodies.**—At night, if the sky be cloudless, we see it spangled with so many STARS that it seems impossible to count them; and we see the same sight in whatever part of the world we may be. The Earth on which we live, is, in fact, surrounded by stars on all sides; and this was so evident to even the first men who studied the heavens that they pictured the Earth standing in the centre of a hollow crystal sphere, in which the stars were fixed like golden nails.

3. In the daytime the scene is changed. In place of thousands of stars, our eyes behold a glorious orb whose rays light up and warm the Earth; and this body we call the SUN. So bright are its beams that, in its presence, all the "lesser lights," the stars, are extinguished. But, if we doubt their being still there, we have only to take a candle from a dark room into the sunshine to understand how

---

1. What is Astronomy? 2. Describe the appearance of the sky at night. By what is the Earth surrounded? 3. In the daytime what do we behold in the sky?

their feeble light, like that of the candle, is "put out" by the greater light of the Sun.

4. There are, however, other bodies which attract our attention. The MOON shines at night, now as a crescent and now as a full Moon, sometimes, like the Sun, rendering the stars invisible. Its changes show us that there is some difference between it and the Sun; for, while the Sun always appears round, because we receive light from all parts of its surface turned toward us, the shape of the bright portion of the Moon varies from night to night, that part only being visible which is turned toward the Sun.

5. Again, if we examine the heavens more closely still, we may see, after a few nights' watching, one, or perhaps two, of the brighter "stars" change their position with regard to the stars lying near them, or relatively to the Sun if we watch that body at its rising and setting. These are the PLANETS; the ancients called them "wandering stars."

6. But the planets are not the only bodies which move across the face of the sky. Sometimes a COMET may by its sudden appearance and strange form awaken our interest, and make us acquainted with a new class of objects, unlike any of those heretofore mentioned.

7. Such are the celestial bodies ordinarily visible. Far away, and comparatively so dim that the naked eye can make little out of them, lie the NEBULÆ (from the Latin *nebula*, "a cloud"); so called because in the telescope they often put on strange cloud-like forms. They differ as much from stars in their appearance as comets do from planets.

There are other bodies, to which we shall refer by and by. We will here merely state in a general way what As-

---

What has become of the stars? 4. What other body do we see at night? What changes does the Moon undergo, and why? 5. On a closer examination of the heavens, what may we see? 6. What bodies sometimes suddenly appear? 7. What other heavenly bodies are mentioned? Describe the nebulæ. 8 What is

tronomy teaches us concerning star and sun, moon and planet, comet and nebula.

8. **The Stars.**—To begin, then, with the stars. So far from being stationary and fixed, as it were, in a hollow glass globe, at nearly equal distances from us, they are all in rapid motion, and their distances vary enormously; though all of them are so very far away that they appear to be at rest, as a ship does when sailing along at a great distance from us. In spite, however, of their great and varying distances, science has been able to get a mental bird's-eye view of all the hosts of stars which the heavens reveal to our eyes, as they would appear to us if we could plant ourselves far on the other side of the most distant one. The telescope—an instrument described further on—has, in fact, taught us that all the stars which we see form but a cluster of islands, as it were, in an infinite ocean of space. We may therefore think of all the stars which we see, as forming *our universe;* and, when we have fixed that thought well in our minds, we may think of space being peopled with *other universes*, as there are other cities besides New York in the United States.

9. Further, we know that our Sun is one of the stars that compose this cluster, and that the reason why it appears so much larger and brighter than the rest is simply because it is the nearest star to us.

We all know how small a distant house looks, or how feebly a distant gas-light seems to shine; but the distant house may be larger than the one we are in, or the distant light be brighter than the one which, being nearer to us, renders the other insignificant. It is precisely so with the stars. Not only would they appear to us as bright as the Sun, if we were as near to them, but we know for a fact that some of them are larger and brighter.

the fact with respect to the motion and distances of the stars? What does the telescope teach us respecting the stars? With what may we regard space as peopled? 9. What is our Sun? Why does it appear larger and brighter than

10. Now, why do the stars and the Sun shine? They shine, or give out light, *because they are white-hot.* They are globes of the fiercest fire; on their surfaces, masses of metals and other substances are burning together more fiercely than any thing we can imagine.

11. **The Planets.**—What, then, are the planets? We may first state that they are comparatively small bodies travelling round our Sun at various distances from him. Our Earth is one of them. There is, however, an important difference between the planets and the Sun. We have seen that the Sun is white-hot; the surface, or outer crust, of our Earth, on the contrary, we know to be cold—all the heat we get coming from the Sun—and because it is cold, it cannot give out light. Astronomers have learned that all the other planets are like the Earth in this respect. They are all dark bodies—having no light in themselves; and they all, like us, get their light and heat from the Sun. When, therefore, we see a planet in the sky, we know that its light is sunshine second-hand; that, as far as its light is concerned, it is but a looking-glass reflecting to us the light of the Sun.

We have now got thus far: *planets are dark or non-luminous bodies travelling round the Sun, which is a bright body—bright because it is white-hot; and the Sun is a star, one of the stars which together form our universe;* the reason that it appears larger and brighter than the other stars being because it is nearer to us than they. It seems likely that the other stars have planets revolving round them, although they are so very far away that the telescopes we possess at present are not powerful enough to show us their planets, if they have any.

12. **The Moon.**—We now come to the Moon. What

---

the other stars? Illustrate this. 10. Why do the stars and the Sun shine? 11. What are the planets? What important difference is there between the planets and the Sun? Whence do the planets get their light and heat? Sum up what we have thus far learned. Are the other stars attended by planets? 12. What

is it? The Moon goes round the Earth, as the planets revolve round the Sun; it is, in fact, a planet of the Earth; it is to the Earth what the Earth is to the Sun. Like the Earth and planets, it is a dark body, and this is the reason it does not always appear round as the Sun does. We only see that part of it that is lit up by the Sun. In the Moon we have a specimen of a third order of bodies, called *satellites*, or companions, as they accompany the planets in their courses round the Sun.

We have, then, to sum up again—(1) The Sun, a star, like all the other stars in motion; (2) Planets revolving round the Sun; (3) Satellites revolving round the planets.

13. **Nebulæ and Comets.**—Nebulæ and comets are very different from the stars and planets, for they are masses of gas. The nebulæ lie far away from us, some of them perhaps out of our universe altogether. The comets rush for the most part from distant regions to our Sun, and having gone round him they go back again, and we only see them for a small part of their journey.

We saw in Art. 10 that the stars shine because they are white-hot; so also nebulæ and comets shine because they are white-hot: but in the case of the stars we are dealing with solid or liquid matter, in the case of the nebulæ and comets with burning gas.

14. **Further Facts.**—Such, then, are some of the bodies with which the science of Astronomy has to deal; but astronomers have not rested content with the appearances of these bodies; they have measured and weighed them, in order to assign to them their true place. Thus they have found that the Sun is 1,245,000 times larger than the Earth, and the Earth 50 times larger than the Moon. They have also discovered that, while we travel round the

---

Is the Moon? Of what order of bodies is it a specimen? 13. How do the nebulæ and comets differ from the stars and planets? Why do they shine? 14. What have astronomers found with respect to the comparative size of the Sun, Earth, and Moon? What, with respect to the distance of the Moon and Sun from the

Sun at a distance of 91,430,000 miles, the Moon travels round us at a distance comparatively insignificant—only 240,000 miles. Thus the greater size of the Sun is balanced, so to speak, by its greater distance; the result being that the large distant Sun looks about the same size as the small near Moon.

15. We already see how enormous are the distances dealt with in Astronomy, although they are measured in the same way as a land-surveyor measures the breadth of a river that he cannot cross. The numbers we obtain when we attempt to measure any distance beyond our own little planetary system convey no impression to the mind. Thus the nearest fixed star is more than 20,000,000,000,000 miles away; the more distant ones are so far away that their light, which travels at the rate of 185,000 miles in a second, requires 50,000 years to reach our eyes!

In spite, however, of this immensity, the methods employed by astronomers are so sure that the distances, sizes, weights, and motions, of the nearer bodies, are now well known. We can, indeed, predict the place that the Moon —the most difficult one to deal with—will occupy ten years hence, with more accuracy than we can observe its position in the telescope.

16. Here we see the utility of the science, and how upon one branch of it, Physical Astronomy, which treats of the motions and structure of the heavenly bodies, is founded another branch, Practical Astronomy, which teaches us how their movements may be made to help mankind.

17. **Usefulness of Astronomy.**—Let us first see what it does for our sailors and travellers. A ship that leaves our shores for a voyage round the world takes with it a book called the "Nautical Almanac," prepared three or four

---

Earth? 15. What kind of distances are dealt with in Astronomy? How far off is the nearest fixed star? How far are the more distant ones? 16. Of what does Physical Astronomy treat? What branch of the science is founded on this?

years in advance by government astronomers. In this book, the places the Moon, Sun, stars, and planets, will occupy at certain stated hours for each day are given, and this information is all that sailors and travellers require to find their way across pathless seas or unknown lands.

But we need not go on board ship or into new countries to find out the practical uses of Astronomy. It is Astronomy that teaches us to measure the flow of time—the length of the day and the year; without Astronomy to regulate them, clocks and watches would be quite useless. It is Astronomy that divides the year into seasons for us, and teaches us the times of the rising and setting of the Moon, which lights up our night. It is to Astronomy that we must appeal when we would inquire into the early history of our planet, or wish to map its surface.

18. Such, then, is Astronomy—the science which, as its name, derived from two Greek words (ἀστήρ, a star, and νόμος, a law), implies, unfolds to us the laws of the stars.

Early History of Astronomy.

19. The establishment of the general facts just stated, and the various laws and principles which constitute the science of Astronomy at the present day, has been the work of centuries. The first astronomers were **the Ancient Shepherds**, who, as they tended their flocks beneath the canopy of heaven, naturally became interested in the orbs with which it was studded, observed their motions, and gave names to those that were most conspicuous. They knew, however, only such isolated facts as were apparent to the eye; it was reserved for later ages to trace visible effects to their causes, and to build up theories; and not till the improved instruments of comparatively recent

---

17. Of what use is Astronomy to sailors? Mention some of the other uses of Astronomy. 18. What is the meaning, and what the derivation, of the word *astronomy?* 19. Who were the first astronomers? What facts alone were

times extended the field of human vision almost beyond belief, was it possible to penetrate the mysteries of the science to their depths.

20. **The Chaldeans** and **Egyptians** were the first to make any material progress in Astronomy. The former, by continued observation, discovered that the eclipses of the Moon recur in the same order in periods of 18 years, and were thus able to predict them with considerable accuracy; the latter investigated the motions of the planets, and established a sacred year of $365\frac{1}{4}$ days.

The Chinese, also, paid great attention to this science in very early times. More than 2,300 years before the Christian era (according to their own records), a tribunal was established for the prosecution of astronomical studies, and particularly for the prediction of eclipses. Its members were held responsible with their lives for the correctness of their calculations; and we are told that one of the emperors actually put to death his two chief astronomers for failing to predict an eclipse of the Sun.

21. From Egypt, the cradle of learning, art, and science, **the Greeks** obtained their first knowledge of astronomy, to which their wise men made important additions. Thales, about 600 B. C., taught that the world was round, and that the Moon shone with reflected light. His pupil Anaximander conceived the bold idea of a plurality of worlds—that is, that the planets are inhabited. A little later, Pythag'oras is said to have advanced the opinion that the Earth and other planets revolve round the Sun. Whether he did so or not, it is certain that this was taught by Aristarchus about 280 B. C.; as, also, that the distance of the Sun from the Earth is insignificant in comparison

---

known to them? 20. What nations were the first to make any material progress in astronomy? What did the Chaldeans discover? What did the Egyptians investigate? What evidence is there that the Chinese paid great attention to astronomy in early times? 21. Whence did the Greeks obtain their knowledge of astronomy? What was taught by Thales? What, by Anaximander? What, by Pythagoras? What, by Aristarchus? What was done by Eratosthenes?

with that of the stars. Among other famous Greek astronomers were Eratos'thenes, who devised an accurate method of measuring the circumference of the Earth, and Hipparchus, who made a catalogue of all the stars visible above his horizon.

Ptolemy, an eminent Egyptian astronomer who flourished in the second century after Christ, rejected the theory of Pythagoras and Aristarchus respecting the solar system, and advanced one of his own, which soon met with general acceptance. He taught that the Earth was the centre of a system of eight immense hollow spheres of crystal, placed one within another: that the Moon was in the nearest sphere; Mercury in the next; Venus in the third; the Sun in the fourth; Mars, Jupiter, and Saturn, in the fifth, sixth, and seventh, respectively; and that the eighth belonged to the stars, which, though most distant, were still visible through the transparent crystal. The revolution of this cumbrous system round the Earth from east to west, once in twenty-four hours, he thought would account for the succession of day and night, and the various phenomena of the heavens.

22. During the Dark Ages, Astronomy was cultivated chiefly by **the Arabians,** who made no advance as regards theory, but were diligent observers, and devised some improvements in instruments and methods of calculation. Even after the termination of this period, comparatively little progress was made until the time of **Coper'nicus,** a Prussian priest, about 350 years ago. He ventured to reject the system of Ptolemy, which then generally prevailed; and, reviving the teachings of Pythagoras and Aristarchus, set forth what is called from him the Copernican system, now very generally received as true, though at first bitterly denounced as visionary and even irreli-

---

What, by Hipparchus? What theory was advanced by Ptolemy? 22. By whom was astronomy chiefly cultivated during the Dark Ages? What improvements were made by the Arabians? When and by whom was the present theory of the

gious. Its three fundamental points are, (1) that the Earth is round; (2) that it turns on its axis from west to east; and (3) that the Earth and other planets revolve round the Sun.

23. After Copernicus came the great Italian philosopher **Galileo**, who first used the telescope, and was thus enabled to make many important discoveries, all tending to support the theory of Copernicus. The day on which Galileo died was memorable for the birth of **Newton**, whose great discovery of the law of gravitation explained the planetary motions, while his mathematical researches gave a new impetus to the science.

Mathematical Definitions.

Certain mathematical terms used in Astronomy must be understood.

24. A Line is a magnitude conceived as having length without breadth or thickness.

25. A Straight Line is one that has the same direction throughout. It is the shortest distance between two points. AB and CD are straight lines.

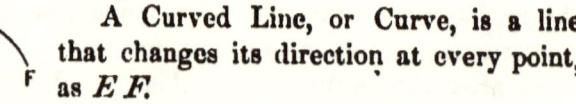

A Curved Line, or Curve, is a line that changes its direction at every point, as EF.

Parallel Lines are such as maintain the same distance from each other at all points, as AB and CD in Figure 1.

26. An Angle is the difference in direction of two straight lines that meet. The point at which they meet is called the Vertex of the angle.

An angle is named from the letter at its vertex, if but

---

universe advanced? What are its three fundamental points? 23. By what two philosophers was Copernicus succeeded, and what discoveries did they make? 24. What is a Line? 25. What is a Straight Line? What is a Curved Line? 26. What is an Angle? How is an angle named? 27. When are two lines said

MATHEMATICAL DEFINITIONS. 19

one angle is formed there. Otherwise, it is named from the letters on each side and at the vertex, that at the vertex being placed in the middle. The angle in Fig. 3 is called K; if more than one angle were formed there, it would be distinguished as IKL or LKI.

Fig. 3.

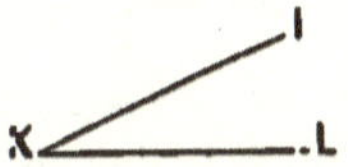

The size of an angle depends not at all on the *length* of its sides, but simply on their *difference of direction*. The angle K will become no larger, however far we may extend its sides.

Fig. 4.

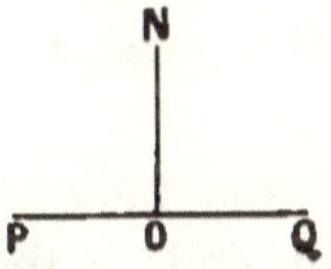

27. When one straight line meets another in such a way as to make the two adjacent angles equal, the lines are said to be *perpendicular* to each other, and the angles formed are called Right Angles. The angles NOP and NOQ, being equal, are right angles; and the lines NO, PQ, are perpendicular to each other.

Fig. 5.

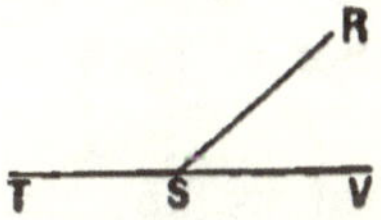

An Obtuse Angle is one that is greater than a right angle, as RST. An Acute Angle is one that is less than a right angle, as RSV.

28. A Surface is a magnitude conceived as having length and breadth without thickness.

A Plane is a surface with which a straight line that joins any two of its points will coincide altogether.

A Convex Surface is one that swells out in a rounded form, as the outside of an egg-shell.

A Concave Surface is one that curves in, as the inside of an egg-shell.

29. A Plane Figure is a plane bounded by a line or lines.

---

to be perpendicular to each other? What are the angles formed by lines that are perpendicular to each other called? What is an Obtuse Angle? What is an Acute Angle? 28. What is a Surface? What is a Plane? What is a Convex Surface? What is a Concave Surface? 29. What is a Plane Figure? 30. What

30. A Triangle is a plane figure bounded by three straight lines, as *X Y Z*.

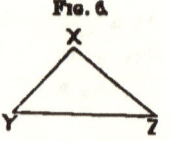

31. A Circle is a plane figure bounded by a curve, every point of which is equally distant from a point within, called the Centre.

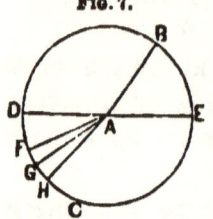

The Circumference of a circle is the line that bounds it. Any part of the circumference is called an Arc. Fig. 7 represents a circle; *A* is the centre, *B E C D* the circumference; *E B*, *B D*, *D F*, etc., are arcs.

A Diameter of a circle is a straight line passing through its centre, and terminating at each end in the circumference, as *D E* in Fig. 7. Every circle has an infinite number of diameters, all equal.

A Radius of a circle is a straight line drawn from the centre to the circumference; as *A B*, *A H*, *A G*, *A F*, in Fig. 7. Every circle has an infinite number of radii, all equal, and each just half its diameter.

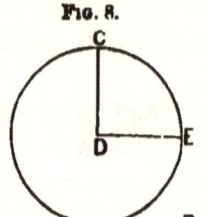

A Tangent is a straight line that touches the circumference of a circle in a single point, without cutting it at either end when produced; as *A B* in Fig. 8.

32. The circumference of every circle may be divided into 360 equal parts, called Degrees (marked °). Each degree may be divided into 60 equal parts, called Minutes ('); and each minute into 60 equal parts, called Seconds (").

A circumference may also be divided into 12 equal parts, of 30 degrees each. These are called Signs.

---

is a Triangle? 31. What is a Circle? What is the Circumference of a circle? What is an Arc? What is a Diameter? What is a Radius? What is a Tangent? 32. Into what may the circumference of every circle be divided? What is a Sign?

MATHEMATICAL DEFINITIONS.

33. A Semicircle is one-half, a Quadrant one-fourth, and a Sextant one-sixth, of a circle.

34. An angle is measured by the number of degrees in the arc that subtends it. A right angle is subtended by one-fourth of the circumference (see CDE in Fig. 8), and is therefore an angle of 90 degrees.

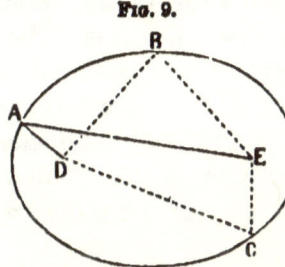

Fig. 9.

35. An Ellipse is a curve every point of which is at such distances from two points within, called its *foci*, that the sum of these distances is in each case the same. ABC is an ellipse; D and E are its foci. $AD + AE = BD + BE = CD + CE.$

An ellipse may be described by fastening the ends of a piece of thread at any two points (the foci) whose distance from each other is less than the length of the thread, and then drawing a line around these points with a pencil placed against the thread, and kept stretched out as far as the thread will allow. The process is shown in Fig. 10.

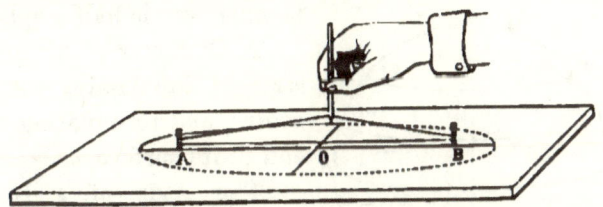

Fig. 10.—MODE OF DESCRIBING AN ELLIPSE.

The thread here represents the sum of the distances from each point of the ellipse to the foci, and remains the same while the ellipse is described. Draw an ellipse according to these directions.

The Centre of an ellipse is the point midway between the foci in the straight line that connects them, as O in Fig. 10. A Diameter is a line passing through the centre and terminating at each end in the ellipse; as GH, IJ.

---

33. What is a Semicircle? What is a Quadrant? What is a Sextant? 34. How is an angle measured? Illustrate this in the case of a right angle. 35. What is an Ellipse? How may an ellipse be drawn? What is the Centre of an ellipse?

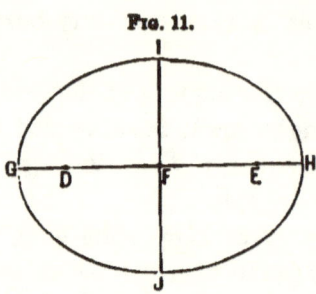

Fig. 11.

The Major Axis of an ellipse is its longest diameter, as *G H*. The Minor Axis is its shortest diameter, as *I J*.

The Eccentricity of an ellipse is the distance of either focus from the centre, divided by half the major axis. Hence the greater the aforesaid distance, the greater the eccentricity, and the more the ellipse deviates from a circle. In Fig. 11, if *D F*, the distance of one of the foci from the centre, is to *G F*, half the major axis, as 2 to 3, the eccentricity of the ellipse will be $\frac{2}{3}$, or .66+.

36. A Solid is a magnitude that has length, breadth, and thickness.

37. A Sphere is a solid bounded by a curved surface every point of which is equally distant from a point within, called the centre. A Hemisphere is half a sphere.

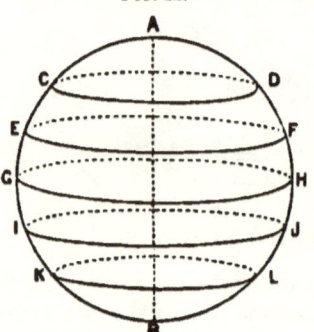

Fig. 12.

A Diameter of a sphere is a straight line passing through its centre, and terminating at each end in its surface.

The Axis of a revolving sphere is the diameter round which it turns. The Poles are the extremities of the axis. In the sphere represented in Fig. 12, the straight line connecting *A* and *B* is a diameter; and also the axis, if the sphere revolves round this diameter; in which case *A* and *B* are the poles.

What is a Diameter? What is the Major Axis? The Minor Axis? What is meant by the Eccentricity of an ellipse? 36. What is a Solid? 37. What is a Sphere? What is a Hemisphere? What is a Diameter of a sphere? What is the Axis of a revolving sphere? What are the Poles? 38. What is a Great

MATHEMATICAL DEFINITIONS.

38. Circles on the surface of a sphere are distinguished as Great and Small. A Great Circle is one whose plane divides the sphere into two equal parts; as *A H B G* and *G H*, in Fig. 12. A Small Circle is one whose plane divides the sphere into two unequal parts, as *C D* and *E F*.

Fig. 13.

The Circumference of a sphere is one of its great circles. The Equator of a sphere is that great circle which is equally distant from the two poles, as *G H* in Fig. 12.

39. A Spheroid is a solid which differs but little from a sphere.

Fig. 14.

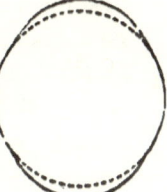

An Oblate Spheroid is a sphere flattened at the poles, as in Fig. 13.

A Prolate Spheroid is a sphere lengthened out at the poles, as in Fig. 14.

CHAPTER I.

THE STARS.

Magnitudes and Distances of the Stars.

40. **Magnitudes of the Stars.**—The first thing that strikes us when we look at the stars is, that they vary very much in brightness. All of those visible to the naked eye are divided into six classes of brightness, called Magnitudes, so that we speak of a very brilliant one as "a

---

Circle? What is a Small Circle? What is the Circumference of a Sphere? The Equator of a Sphere? 39. What is a Spheroid? An Oblate Spheroid? A Prolate Spheroid?

40. What is the first thing that strikes us when we look at the stars? As re-

star of the first magnitude:" of the feeblest visible, as "a star of the sixth magnitude," and so on.

The number of stars of all magnitudes visible to the naked eye under the most favorable circumstances, is about 6,000; so that the greatest number visible at any one time—as we can only see half of the sky at once—is 3,000. If we use a small telescope this number is largely increased, as that instrument enables us to see stars too feeble to be perceived by the eye alone. The stars thus revealed to us are called Telescopic Stars. These also vary in brightness; and the classification is continued down to the twelfth, fourteenth, sixteenth, and even higher magnitudes, according to the power of the telescope. With powerful telescopes, at least 20,000,000 stars down to the fourteenth magnitude are visible.

41. **Comparative Brightness.**—A star of the 6th magnitude is, as we have seen, the faintest visible to the naked eye. Taking the average brightness of a 6th-magnitude star as unity, the average brightness of the other classes is estimated as follows:—

6th magnitude,............1	3d magnitude,............12
5th " 2	2d " 25
4th " 6	1st " 100
Sirius, the brightest star of the 1st magnitude,...............324
The Sun, the nearest star to us,................6,480,000,000,000

Even stars of the same magnitude differ considerably in brilliancy. It will be seen from the above table that the brightness of Sirius is more than three times as great as the average brightness of its class.

42. The stars are usually divided about as follows: of the first magnitude, 20; of the second, 65; of the third,

---

gards brightness, how are the stars visible to the naked eye divided? How many are there? How many can be seen at once? What are Telescopic Stars? How are they divided? How many stars are there, including those of the 14th magnitude? 41. What is the comparative brightness of the stars of the first six magnitudes? How does the brightness of Sirius compare with the average brightness of its class? How does the Sun compare in brightness with a star of the 6th magnitude? 42. State the number of stars of each of the first six magnitudes.

200; of the fourth, 450; of the fifth, 1,100; of the sixth, about 4,000. The number increases largely as we descend in the scale of brilliancy.

43. **Distances of the Stars.**—It is evident that the stars, as they shine with such different lights, one star differing from another star in glory, are either of the same size at very different distances, the most remote being of course the faintest; or are of different sizes at the same distance, the largest shining the brightest; or are of different sizes at different distances. In the case of twelve stars the actual distances are known, and differ greatly; as regards the rest, we can only say it is most probable that the difference in brilliancy is due mainly to difference of distance.

44. The distances of the stars from us are so great that to state them in miles hardly gives an adequate idea of them; some other method, therefore, must be used, and the velocity of light affords us a convenient one.

Light travels at the rate of 185,000 miles in a second— that is to say, between the beats of the pendulum of an ordinary clock, light travels a distance equal to eight times round the Earth. Now, the nearest star (leaving the Sun out of the question) is situated at a distance which light, even with the extraordinary velocity just mentioned, requires three and a half years to traverse. We may say that, on an average, light requires fifteen and a half years to reach us from a star of the 1st magnitude, twenty-eight years from a star of the 2d, forty-three years from a star of the 3d, one hundred and twenty from a star of the 6th, and so on, until for stars of the 12th magnitude the time required is thirty-five hundred years. If, therefore, a star of the 6th magnitude were destroyed at the present moment, we should continue to see it in the heavens for 120 years to come; and if one of the 12th magnitude

---

43. To what are the different degrees of brightness in the stars due? 44. Give an idea of the distances of the stars, as measured by the velocity of light.

were now created, it would be 3,500 years before it would be perceptible to us.

45. **The Diameters of the Stars** cannot be determined by our most powerful instruments. As seen from the Earth, they are, in consequence of their distance, mere points of light, so small as to be beyond our most delicate measurement. The Moon, which travels very slowly across the sky, sometimes gets before, or eclipses, or occults, some of them; but they vanish in a moment—which they would not do, if they were not as small as we have stated.

46. **The Milky Way.**—Winding among the stars, a beautiful belt of pale light spans the sky, and sometimes it is so situated as to divide the heavens into two nearly equal portions. This belt is the Milky Way (see Celestial Charts at the end of the volume); and the smallest telescope shows that it is composed of stars so faint, and apparently so near together, that the eye can perceive only a dim continuous glimmer. Milton alludes to it as the

"Broad and ample road
Whose dust is gold, and pavement stars."

Among the Greeks, the Milky Way was known as the Galaxy and the Circle of Milk. The Chinese and Arabians call it the Celestial River. Some of the American Indian tribes regarded it as the path of departed souls to the spirit-land. In England it used to be familiarly called Jacob's Ladder.

Different opinions prevailed among the ancients as to what it was. Aristotle thought it was the result of gaseous exhalations from the earth set on fire in the sky. Theophrastus believed it to be the soldering together of two hemispheres constituting the celestial vault. Diodo'rus represented it as a dense celestial fire, appearing through the clefts of parting hemispheres. Democ'ritus and Pythagoras divined the truth, that the Galaxy is nothing more or less than a vast assemblage of very distant stars; and Ovid speaks of it as a highway whose groundwork is of stars.

---

45. What is said of the size or diameters of the stars? What happens when the Moon eclipses one of them? 46. What is the Milky Way? In what terms does Milton allude to it? What did the Greeks call the Milky Way? The Chinese and Arabians? How did some of the American Indians regard it? What did it use to be called in England? Give some of the views of the ancients respecting

THE MAGELLANIC CLOUDS.

47. **The Magellanic Clouds,** called the *Nubec'ula Major* and *Nubec'ula Minor*, distinctly visible in the southern hemisphere on a clear moonless night, are two cloudy oval masses of light, very like portions of the Milky Way, but apparently unconnected with its general structure. The telescope resolves them into single stars, star-clusters, and nebulous matter in various degrees of condensation. They are named from the Portuguese navigator Magellan, though he was not the first to observe them.

Fig. 15 shows the appearance which part of the *Nubecula Major* presents, when viewed through the telescope.

FIG. 15.—PART OF THE NUBECULA MAJOR.

Shape of our Universe.

48. **Distribution of the Stars.**—The largest stars are scattered very irregularly; but, if we look at the smaller ones, we find that they gradually increase in number as they approach the Milky Way. In fact, of the 20,000,000 stars visible, as we have stated, in powerful telescopes, at least 18,000,000 lie in and near the Milky Way.

---

the Milky Way. 47. Describe the Magellanic Clouds. Into what does the telescope resolve them? 48. How do we find the largest stars distributed? How.

49. Adding this fact to what has been said about the distances of the stars, we can now determine the shape of our universe. It is clear that it is most extended where the faintest stars are visible, and where they appear nearest together; because they appear faint in consequence of their distance, and because their apparent close packing arises, not from actual nearness to each other, but from their lying in that direction at constantly increasing distances. Indeed, the stars which form the Milky Way, extending one behind another to an almost infinite distance, are probably as far from each other as our Sun is from the nearest star.

50. The Milky Way, then, traces for us the direction in which our universe has its largest dimensions. The absence of faint stars in the parts of the sky farthest from the Milky Way shows us that the limits of the universe in that direction are much sooner reached than in the direction of the Milky Way itself. We gather, therefore, that the thickness of our universe is small compared with its length and breadth; that its shape is not spherical, but rather that of a circular piece of thick pasteboard. And as the Milky

FIG. 16.—SUPPOSED SHAPE OF OUR UNIVERSE.

---

the smaller ones? 49. What may be inferred respecting the shape of our universe? 50. What does the Milky Way trace for us? What does the absence of faint stars in the parts of the sky remote from the Milky Way show? What is

Way divides into two principal branches, which, after pursuing separate courses nearly half its length, again unite, we infer that this flattened stratum of stars is divided lengthwise, as if the rim of the pasteboard were split and its two surfaces pulled apart at a small angle through half the circle, as in Fig. 16.

To account for the appearances presented, we must regard our solar system as lying nearly at the centre of this mass of stars, and near the region at which it begins to divide; but, as there are more stars on the south side of the Milky Way than there are on the north, we gather that our Earth occupies a position somewhat to the north of the middle of its thickness.

On this supposition, all the stars which, owing to our position in observing them, appear so remote from the Milky Way, really form part of it, and our great Sun represents but an atom of its luminous sand.

51. Although the Milky Way thus enables us to get a rough idea of the shape of our universe, as we get an idea of the shape of a wood from some point within it by seeing in which direction the trees appear thickest together, still the telescope teaches us that its boundaries are probably very irregular.

The Constellations.

52. We have thus far considered our star-system as a whole, its dimensions, and shape. Before we proceed with a detailed examination of the stars composing it, it will be convenient to state the groupings into which they have been arranged, and the way in which any particular star may be referred to.

53. The stars, then, from remote antiquity, have been classified into groups called **Constellations**, each constella-

---

Inferred from the fact that the Milky Way divides into two principal branches? Where must we regard our solar system as lying? 51. What does the telescope teach us respecting the boundaries of the Milky Way? 53. How have the stars

tion being fancifully named after some object (in most cases, an animal or mythological personage) which the arrangement of the stars composing it was thought to suggest. For the most part, however, little or no resemblance can be traced to the object after which the group is named; compare, for instance, the outline of the Great Bear with the position of the principal stars composing that constellation, as shown in Fig. 17.

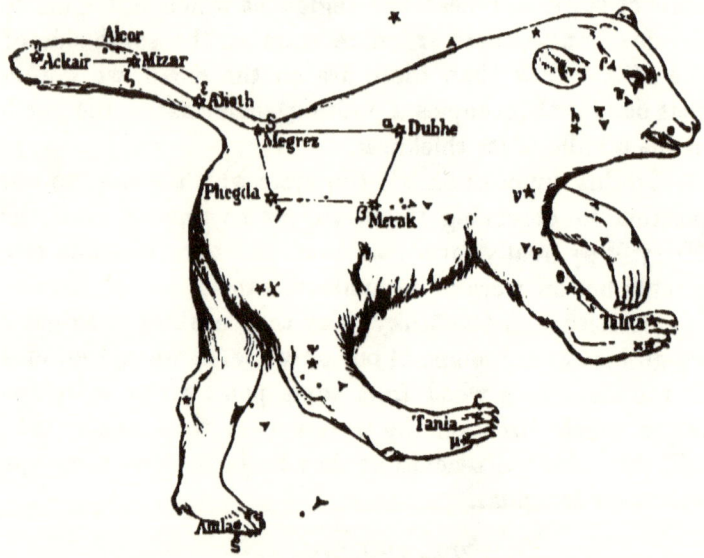

FIG. 17.—CONSTELLATION OF THE GREAT BEAR.

The resemblance being in most cases so remote that the effort to trace the figures in the sky is unsatisfactory and confusing, it has been thought better to present, in the Celestial Charts at the close of this volume, the boundaries and relative positions of the constellations, as indicated by dotted lines, than to delineate the animals and fabulous personages from which they are named.

54. Some of the most marked constellations probably received their names 1,500 years before the Christian era, and all the leading ones were known in the time of Ara'-

---

from remote antiquity been classified? After what are the constellations named?
54. How early were the leading constellations known? Who added two in the

tus (270 B. C.), who described them in verse. About 150 years after Christ, Ptolemy arranged in 48 constellations the 1,022 stars which Hipparchus had observed at Rhodes 250 years before. Tycho Brahe, a Danish astronomer, added two constellations in the sixteenth century; and to these 50 (called the *ancient* constellations), 59 have since been added, making the present number 109.

55. **The Zodiacal Constellations.**—The Latin and English names of the ancient constellations and the most important modern ones will now be given. We begin with the twelve through which the Sun passes in his annual round. These are called the Zodi'acal Constellations; they are to be very carefully distinguished from the *signs of the zo'diac* bearing the same name. Their names should be learned in the order in which they are presented, and should be found on the Celestial Charts at the end of the volume, where the constellations are laid down in order on the heavy circle called the Ecliptic:—

A'ries,	the Ram.	*Li'bra,*	the Balance.
Taurus,	the Bull	*Scorpio,*	the Scorpion.
Gem'ini,	the Twins.	*Sagitta'rius,*	the Archer.
Cancer,	the Crab.	*Capricornus,*	the Goat.
Leo,	the Lion.	*Aquarius,*	the Water-bearer.
Virgo,	the Virgin.	*Pisces,*	the Fishes.

The order of the names may perhaps be more readily remembered, if they are thrown together in rhyme:—

>The Ram and Bull lead off the line;
>Next Twins, and Crab, and Lion, shine,
>The Virgin and the Scales;
>Scorpion and Archer next are due,
>The Goat and Water-bearer too,
>And Fish with glittering tails.

56. **The Northern Constellations** are those which are

16th century? How many have been added in modern times? What is the present number? 55. What is meant by the Zodiacal Constellations? From what

visible above the zodiacal constellations. The principal ones are as follows (see Celestial Chart of the Northern Hemisphere):—

Ursa Major,	the Great Bear.
Ursa Minor,	the Little Bear.
Draco,	the Dragon.
Cepheus,	Cepheus.
Boötes,	Boötes.
Coro'na Borea'lis,	the Northern Crown.
Hercules,	Hercules.
Lyra,	the Lyre.
Cygnus,	the Swan.
Cassiope'a,	Cassiopea (the Lady's Chair).
Perseus,	Perseus.
Auri'ga,	the Wagoner.
Ophiuchus,	the Serpent-bearer.
Serpens,	the Serpent.
Sagitta,	the Arrow.
A'quila,	the Eagle.
Delphi'nus,	the Dolphin.
Equuleus,	the Little Horse.
Peg'asus,	the Winged Horse.
Androm'eda,	Andromeda.
Triangulum,	the Triangle.
Camelopardalus,	the Camelopard.
Canes Venat'ici,	the Hunting Dogs.
Coma Bereni'ces,	Berenice's Hair.
Vulpec'ula et Anser,	the Fox and the Goose.
Cor Car'oli,	Charles's Heart.

57. **The Southern Constellations.**—The principal constellations visible in the United States below the zodiacal ones, called Southern Constellations, are as follows (see Celestial Chart of the Southern Hemisphere):—

---

are they to be distinguished? Name the Zodiacal Constellations in order. 56. Name some of the principal Northern Constellations. 57. Name some Southern

Cetus,	the Whale.
Ori'on,	Orion.
Erid'anus,	the River Eridanus.
Lepus,	the Hare.
Canis Major,	the Great Dog.
Canis Minor,	the Little Dog.
Argo Navis,	the Ship Argo.
Hydra,	the Snake.
Crater,	the Cup.
Corvus,	the Crow.
Centaurus,	the Centaur.
Lupus,	the Wolf.
Coro'na Austra'lis,	the Southern Crown.
Piscis Australis,	the Southern Fish.
Monoc'eros,	the Unicorn.
Columba Noachi,	Noah's Dove.

58. **Names of the Stars.**—The whole heavens being portioned out among these constellations, the next thing to be done was to invent some method of referring to each particular star. The method introduced by John Bayer, of Augsburg, in 1603, and now in use, is to arrange all the stars in each constellation in the order of brightness, and to attach to them in that order the letters of the Greek alphabet, using after the letters the genitive of the Latin name of the constellation. Thus, *Alpha* (α) *Lyræ* denotes the brightest star in the Lyre; *Beta* (β) *Ursæ Minoris*, the next to the brightest star in the Little Bear.

Except, however, in the case of the brighter stars of a constellation, this alphabetical arrangement has not been strictly adhered to, and consequently it does not always indicate the relative brilliancy of the less important stars.

59. The letters of the Greek alphabet, and their names, are as follows:—

Constellations visible in the United States. 58. Describe and illustrate the method of referring to particular stars.. 59. Repeat the Greek alphabet. After

34 · THE STARS.

α,	Alpha.	ι,	Iota.	ρ,	Rho.
β,	Beta.	κ,	Kappa.	σ,	Sigma.
γ,	Gamma.	λ,	Lambda.	τ,	Tau.
δ,	Delta.	μ,	Mu.	υ,	Upsilon.
ε,	Epsilon.	ν,	Nu.	φ,	Phi.
ζ,	Zeta.	ξ,	Xi.	χ,	Chi.
η,	Eta.	ο,	Omicron.	ψ,	Psi.
θ,	Theta.	π,	Pi.	ω,	Omega.

After the Greek alphabet is exhausted, the Roman alphabet is used in the same way; and after that recourse is had to numbers.

60. Some of the brightest stars are still called by the Arabian or other names by which they were formerly known. Thus, *a Canis Majoris* is known also as Sir'ius; *a Boötis*, as Arctu'rus; *β Orionis*, as Rigel; *a Lyræ*, as Vega; *a Tauri*, as Aldeb'aran; *a Ursæ Minoris*, as Pola'ris (the Pole-star), etc.

61. The constellations that have been named, and their principal stars, can be seen on the charts at the end of this book, or on a Celestial Globe. Before proceeding further, the student should make himself familiar with them, that he may know their relative positions when he comes to trace them in the sky.

In star-maps the stars are laid down as we actually see them in the heavens, looking at them from the Earth; but in globes their positions are reversed, as the Earth, on which the spectator is placed, is supposed to occupy the centre of the globe, while we really look at the globe from the outside. If, therefore, the brighter of two stars appears on the right of the other in the heavens, it will be shown on its right in a star-map, but to the left of it on a globe.

62. **Stars of the First Magnitude.**—The twenty brightest stars in the heavens, or first-magnitude stars, are as fol-

---

the Greek alphabet is exhausted, what are used in naming the stars? 60. What other names have some of the brightest stars? Give examples. 61. How are the stars laid down in star-maps? How, on globes? Name the twenty brightest

lows; they are given in the order of brightness, and should be found on the charts or on a globe:—

Sirius,	in Canis Major.	Aldeb'aran,	in Taurus.
Cano'pus,	" Argo Navis.	Beta,	" Centaurus.
Alpha,	" Centaurus.	Alpha,	" Crux.
Arcturus,	" Boötes.	Anta'res,	" Scorpio.
Rigel,	" Orion.	Altair,	" Aquila.
Capella,	" Auriga.	Spica,	" Virgo.
Vega,	" Lyra.	Fomalhaut,	" Piscis Australis.
Pro'cyon,	" Canis Minor.	Beta,	" Crux.
Betelgeuse	" Orion.	Pollux,	" Gemini.
Achernar,	' Eridanus.	Regulus,	" Leo.

Motions of the Stars.

63. **Proper Motion.**—Now, although the stars and constellations retain the same relative positions as they did in ancient times, all the stars are, nevertheless, in motion; and in some of those nearest to us, this motion, called Proper Motion, is very apparent, and has been measured. Thus Arcturus is travelling at the rate of at least fifty-four miles a second, or three times faster than our Earth travels round the Sun—or six thousand times faster than an ordinary railway train.

64. Nor is our Sun, which be it remembered is a star, an exception; it is approaching the constellation Hercules at the rate of four miles in a second, carrying its system of planets, including our Earth, with it. Here, then, we have an additional cause for a gradual change in the positions of the stars, for a reason we shall readily understand, if, when we walk along a gas-lit street, we notice the distant lights. We shall find that the lights we leave behind close up, and those in front of us open out as we approach them: so the stars which our system is approaching are

---

stars, and state what constellation each is in. 63. What is meant by the Proper Motion of the stars? At what rate is Arcturus travelling? 64. What additional cause have we for a gradual change in the positions of the stars? How is this illustrated in the case of a gas-lit street? 65. From what motions are the Proper

slowly opening out, while those we are quitting are closing up, as our distance from them increases.

65. **Apparent Motion.**—The real motions of the stars—called, as we have seen, their proper motions—and the one we have just pointed out, are, however, to be gathered only from the most careful observation, made with the most accurate instruments. There are Apparent Motions, which may be detected in half an hour by the most careless observer. These are caused, as we shall fully explain in Chap. XII., by the two real motions of the Earth, first round its own axis, and secondly round the Sun..

Double and Multiple Stars.

66. An examination of the stars with a powerful telescope, reveals to us the most startling and beautiful appearances. Stars which appear single to the unassisted eye, appear double, triple, and quadruple; and in some instances the number of stars revolving round a common centre is even greater. Because our Sun is an isolated star, and because the planets are now dark bodies, instead of shining, like the Sun, by their own light, as they once must have done, it is difficult, at first, to realize such phenomena, but they are among the most firmly-established facts of modern astronomy.

FIG. 18.—ORBIT OF A DOUBLE STAR.

A beautiful star in the constellation Lyra will at once give an idea of such a system, and of the use of the telescope in these inquiries. The star in question, *Epsilon* (ε) *Lyræ*, to the naked eye appears as a faint single star. A

Motions to be distinguished? By what are the Apparent Motions caused? 66. What changes in the appearance of some stars does a powerful telescope produce? Give an account of *Epsilon Lyræ*. In what times will the members of this

DOUBLE AND MULTIPLE STARS.

small telescope, or opera-glass even, suffices to show it double, and a powerful instrument reveals the fact that each star composing this double is itself double; hence it is known as "the Double-double." Here, then, we have a system of four stars; the stars composing each pair, considered by themselves, revolving round a point between them; while the two pairs, considered as two single stars, perform a much larger journey round a point situated between them.

It may be stated roundly that the wider pair will complete a revolution in 2,000 years; the closer one, in half that time; and possibly both double systems may revolve round the point lying between them in something less than a million of years.

FIG. 19.—THE DOUBLE-DOUBLE STAR IN THE CONSTELLATION LYRA. 1. As seen in an opera-glass. 2. As seen in a small telescope. 3. As seen in a telescope of great power.

67. Of the multiple stars, that is, such as are resolved by the telescope into more than four single stars, *Theta* (θ) *Orionis* is one of the most interesting. What appears to the unaided eye as a single luminous point, is shown by a powerful telescope to consist of seven stars, arranged as shown in Fig. 20.

FIG. 20.—THE MULTIPLE STAR, θ *Orionis*.

68. More than 6,000 double stars are now known, in nearly 700 of which a regular orbital motion, and in some of them a very rapid motion, has already been detected.

---

system complete their revolution? 67. Describe *Theta Orionis*, as seen through a powerful telescope. 68. How many double stars are now known? In how many of these has an orbital motion been detected? How do the double stars

In some cases the brilliancy of the component stars is nearly equal; in others the light is very unequal. For instance, a first-magnitude star may have a companion of the fourteenth magnitude. Sirius has at least one such companion. Here is a list of some double stars, showing the time in which a complete revolution is effected:—

	Years.		Years.
Zeta (ζ) Herculis,	36	Gamma (γ) Coronæ Borealis,	100
Eta (η) Coronæ Borealis,	43		
Zeta (ζ) Cancri,	60	Delta (δ) Cygni,	178
Alpha (α) Centauri,	75	Beta (β) Cygni,	500
Omega (ω) Leonis,	82	Gamma (γ) Leonis,	1200

69. In the case, then, of nearly 700 double stars, in which an orbital motion of the component stars round a common centre of gravity has been observed, there can be no doubt that we have connected systems. Pairs thus connected are called Binary Stars, or *Physical Couples*, to distinguish them from unconnected double stars, or *Optical Couples*, in which the component stars are really distant from each other, their apparent nearness being due to their lying in the same straight line as seen from the Earth.

70. When the distance of a binary star is known, we can determine the dimensions of the orbit of one star round the other, as we can determine the Earth's orbit round the Sun. Thus, in the case of the binary star 61 *Cygni*, its distance from us being known, it is found that the orbit of the smaller of the two stars has a mean radius of about 45 times the distance of the Earth from the Sun, or more than 4,275,000,000 miles. And yet so immense is the distance of the two stars from us, that to the naked eye they seem as one.

---

differ, as regards the comparative brightness of the individual stars composing them? Mention some of the double stars, with their period of revolution, and point to them on the Celestial Chart. 69. What are Binary Stars, or Physical Couples? From what must they be distinguished? 70. When the distance of a binary star is known, what can be determined? What is the size of the orbit

Variable Stars.

71. Variations in Brightness.—The stars are not only of different magnitudes (Art. 40), but the brilliancy of some particular stars changes from time to time. Stars whose brightness varies slowly, regularly, and within certain limits, are called Variable Stars, or briefly Variables. In some few cases, however, the increase and decrease have been sudden, and in others the limits of change are unknown; hence we read of new stars, lost stars, and temporary stars, in addition to the more regular variables. There is little doubt, however, that all these phenomena are the same in kind, though different in degree.

72. Amount and Period of Variation.—The variation in brightness is measured by the difference between the greatest and the least magnitude of the star at different times. The interval between two successive times when the star is brightest is called the Period of Variation.

73. Table of Variable Stars.—There are more than 100 variable stars whose periods are known, besides others whose periods have not been determined. The following table contains a few of the former class:—

	Change of Magnitude, from	to	Period of Variation.
η *Argûs*,	1	4	46 years.
R *Cephei*,	5	11	73 "
R *Cassiopeæ*,	6	lower than 14	435 days.
ο *Ceti*,	1 or 2	lower than 11	$331\frac{1}{3}$ "
ς *Cancri*,	8	$10\frac{1}{4}$	10 "
β *Persei*,	$2\frac{1}{2}$	4	$2\frac{9}{10}$ "

74. Mira.—The fourth star in the above table, called also *Mira*, or "the marvellous," has been known as a variable for about three centuries. It preserves its greatest

_{of the smaller of the two stars in 61 *Cygni?* 71. What is meant by Variable Stars? When the increase and decrease have been sudden, what have variables been called? 72. By what is the variation in brightness measured? What is meant by the Period of Variation? 73. How many variable stars are there, whose periods are known? Mention one or two, with their change of magnitude}

brilliancy for about fifteen days, generally appearing at that time as a star of the first or second magnitude, but occasionally not brighter than one of the fourth. For the next three months its light decreases till it becomes invisible, not only to the naked eye, but even in telescopes of small power. It so remains for five months, then reappears, and in about three months again attains its maximum brightness, to repeat the same phases. Irregularities have been discovered in its period of variation, but these irregularities are themselves periodical.

75. **Algol.**—Among the variables, *Beta* (β) *Persei*, or *Algol*, is perhaps the most interesting, as its period is short, and it never becomes invisible to the naked eye. It shines as a star of the second magnitude for two days thirteen hours and a half, and then suddenly loses its light, and in three hours and a half falls to the fourth magnitude; its brilliancy then increases again, and in another period of three hours and a half it reattains its greatest brightness —all the changes being accomplished in less than three days.

76. **New, or Temporary, Stars.**—Among the New, or Temporary, Stars, those observed in 1572 and 1866 are the most noticeable. The first appeared suddenly in the sky and was visible for seventeen months. Its light at first was equal to that of the planets at their greatest brilliancy; so bright was it, indeed, that it was clearly visible at noonday. Now, it is not a little curious that in the years 945 and 1264 something similar was observed in the same region of the sky (in Cassiopea) in which this star appeared. If, then, we assume that we have here a variable star of long period, which is very bright at its maximum and fades out of view at its minimum, we may expect a reappearance of the star about the year 1885.

---

and period of variation. 74. Give an account of Mira. 75. Describe the changes of Algol. Point to Mira and Algol on the Celestial Charts. 76. Which are the most noticeable of the New, or Temporary, Stars? Give an account of the one

CAUSE OF VARIATION IN BRIGHTNESS.

We now come to the new star which broke upon our sight in 1866, in the constellation Corona Borealis, and which was observed with powerful methods of research not employed before. This star was recorded some years ago as one of the ninth magnitude. In May, however, it suddenly flashed up, and on the 12th of that month shone as a star of the second magnitude. On the 14th it descended to the third magnitude; the decrease of brightness was for some time at the rate of about half a magnitude a day, but toward the end of May it was less rapid. There is good reason to believe that this increased brilliancy was due to the sudden ignition of hydrogen gas in the star's atmosphere. Here we have a fact of the highest importance, though as yet we can hardly do more than speculate upon it.

77. **Cause of Variation.**—The cause of this change of brightness in variable stars is one of the most puzzling questions in the whole domain of Astronomy. Three theories have been advanced:—

1. That the variable revolves on its axis; that its surface is not equally luminous in all parts; and hence that it appears more or less bright, according to the part that is presented toward us.

2. That the variable is accompanied by non-luminous planets, which in the course of their revolution get between us and the variable, and thus eclipse the latter either in whole or in part.

3. The most recent theory is that of Balfour Stewart, deduced from his researches on the Sun, which is doubtless a variable star. He has found that the approach of a planet to our Sun increases its brightness, especially in that part which is nearest to the planet. Hence he supposes that the variable has a large planet revolving round

---

that appeared in 1572. Describe the new star that appeared in 1866. To what is the increased brilliancy of this star attributed? 77. What three theories have been advanced, to account for the change of brightness in variable stars? 78.

it at a small distance; that part of the star which is nearest the planet will then be more luminous than that which is more remote, and, as the planet revolves, an appearance of variation, with a period equal to that of the planet's revolution, will be presented to the observer.

If we suppose the planet to have a very elliptical orbit, then for a long time it will be at a great distance from its primary, while for a comparatively short time it will be very near. We should, therefore, expect a long period of darkness, and a comparatively short one of intense light—precisely what we have in temporary stars.

Colored Stars.

78. The light of most of the stars is white; but there are a number of **Colored Stars**, which shine with red, orange, purple, blue, or green light,—some but faintly tinged, and others having a very deep and decided hue. Of large stars of different colors we may give the following table:—

Red . . *Aldeb'aran, Anta'res, Betelgeuse.*
Blue . . *Capella, Rigel, Bella'trix, Pro'cyon, Spica.*
Green . *Sirius, Vega, Altair, Deneb.*
Yellow . *Arctu'rus.*
White . *Reg'ulus, Denebola, Fomalhaut, Polaris.*

79. **Colored Double Stars.**—It is in the double and multiple stars that the richest colors are presented; and in these we also frequently find striking contrasts. Thus, in the double star *Iota* (ι) *Cancri*, the larger of the two is orange, the smaller blue. The triple star *Gamma* (γ) *Andromedæ* is formed of an orange-red sun, accompanied by two others of an emerald green. In *Eta* (η) *Cassiopeæ* we have a large white star, with a companion of a rich ruddy purple.

What color is the light of most stars? What is the color of some? Mention some of the large colored stars. 79. In what stars are the richest colors pre-

COLORED STARS. 43

What wondrous coloring must be met with in the planets lit up by these glorious suns, one sun setting, say in clearest green, another rising in purple, or yellow, or crimson; at times two suns at once mingling their variously-colored beams! A remarkable group in the Southern Cross produced on Sir John Herschel "the effect of a superb piece of fancy jewelry." It is composed of over 100 stars, only seven of which exceed the tenth magnitude; two of this group are red, two green, three pale green, and one greenish blue.

80. **Changes of Color.**—In many cases, the colors of the stars have changed. If we go back to ancient times, we read that Sirius was fiery-red; it gradually faded to a pure white, and is now a decided green. Capella was also described as red; it afterward became yellow, and is now a pale blue.

In some variable stars the changes of color are very striking. In the new star of 1572, Tycho Brahe observed changes from white to yellow, and then to red; and we may add that generally, when the brightness decreases, the star becomes redder.

81. The variations which the stars undergo in brightness and color, while we can not yet speak with certainty as to their causes, indicate that incessant movement and change are going on in the distant regions of space.

Structure of the Stars.

82. **The Photosphere.**—We will now pass on to what is known of the physical constitution of the stars. In the first place, the stars, of whatever their interiors may be composed, present to us on their exteriors a bright surface, which is called the Photosphere; outside of this

---

sented? Give examples. What remarkable group in the Southern Cross is mentioned? 80. State some remarkable changes of color. In what stars are changes of color very striking? 81. What do the changes in the brightness and color of the stars indicate? 82. What is the Photosphere of a star? What is

photosphere, as outside the surface of our Earth, is an atmosphere composed of vapors. The materials of the photospheres are intensely hot; so hot, that the metals and other substances of which they consist are in a liquid or vaporous state.

We can render this intelligible by taking water and iron as examples. When both are in a solid state, we have ice and hard iron. If we apply heat, we melt both ice and iron; but we find that it requires much more heat to convert the latter into a liquid state than it does the former—the melting-point of ice being $32°$ of Fahrenheit's thermometer, while that of iron is about $2,000°$. Having reduced both to liquids, we may by additional heat turn the water into steam, and the molten iron into iron-vapor; but again the heat needed to vaporize the iron is vastly greater than that required to vaporize the water—how much greater is not known, as the heat necessary to produce iron-vapor exceeds our powers of measurement. So, also, does the heat present in the photospheres of the stars.

83. **Materials of the Photospheres.**—Do we know any thing of the substances which throw out this heat and light? Yes, a little. For instance:—

Sirius contains sodium, magnesium, iron, and hydrogen.
Vega " sodium, magnesium, and iron.
Pollux " sodium, magnesium, and iron.
Beta Pegasi " sodium, magnesium, and perhaps barium.

It is remarkable that the elements most widely diffused among the stars, including hydrogen, sodium, magnesium, and iron, are among those most closely connected with the living organisms of our globe.

We shall be able, when we come to examine the

---

outside of this photosphere? What is the temperature of the materials of the photosphere? Show the effect of heat, by taking water and iron as examples. 83. State the materials which the photospheres of several of the brightest stars are found to contain. What is remarkable with respect to these elements? 84.

structure of the nearest star—the Sun—to obtain a more detailed knowledge of the structure of the stars generally.

84. **Causes of Color in the Stars.**—The vapors produced in the photospheres of the stars ascend to form atmospheres, and these atmospheres absorb, in part, the light given out by the photospheres. A piece of colored glass will teach us what is meant by the absorption of light, and how it produces color. Thus, green glass is green because it absorbs all other light but the green; it is a sort of sieve, which stops every ray of light except the green ones. So with glasses, solids, vapors, or liquids, of other colors.

Now, the colors of the stars may be influenced, not only by the degree of heat in their photospheres, but by the amount of absorption in their atmospheres. Our Sun at setting, for instance, sometimes seems blood-red, in consequence of the absorption of *our* atmosphere; if the absorption were in his own atmosphere, he would be blood-red at noonday.

Concerning the causes which produce the *changes* in color and brightness. we must confess that, after all, we are yet ignorant.

Star Groups and Clusters.

85. Having now dealt with the peculiarities of individual stars,—their distance, arrangement, color, variability, and structure,—we next come to the various assemblages of stars observed in various parts of the heavens.

86. **Remarkable Star-groups.**—In the double and multiple systems (Art. 66) we saw the first beginnings of the tendency of the stars to group themselves together. In some parts of our system this tendency is exhibited in a very remarkable manner, the beautiful group of the

---

How is the color of the stars explained? What is said of the cause of changes of color? 86. What tendency is seen in the double and multiple systems of

Pleiades (which may be found on the Celestial Chart, in the constellation Taurus) affording a familiar instance. The six or seven stars visible to the naked eye become 60 or 70 when viewed in the telescope. The Hyades (near the Pleiades, in Taurus), and Præsepe, or "the Beehive," in Cancer, may also be mentioned.

In other cases, the groups consist of an innumerable number of suns apparently closely packed together. That in the constellation Perseus appears like a nebula to the naked eye, but viewed through a telescope it is separated into stars, and forms one of the most beautiful objects in the heavens. Many others, scarcely less stupendous, though much fainter by reason of their greater distance, are revealed by the telescope.

87. Assemblages of stars are divided into,
   1. **Irregular Groups,** generally more or less visible to the naked eye.
   2. **Star-clusters,** invisible to the naked eye, but which, in the most powerful telescopes, are seen to consist of separate stars. These are subdivided into Ordinary Clusters and Globular Clusters.

Clusters and nebulæ are designated by their number in the catalogues which have been made of them by different astronomers. The most important of these catalogues are those of Messier, Sir William Herschel, and Sir John Herschel. About 5,400 nebulæ have been observed.

88. Of the Ordinary Star-clusters, the magnificent ones in the constellations Libra and Hercules (represented in Figs. 1 and 2, on page 47) may be mentioned as among those which are best seen in telescopes of moderate power. The Globular Clusters are well represented by those in Serpens and Aquarius (see Figs. 4 and 5, p. 47).

89. **Other Universes.**—Some of the clusters which lie

ORDINARY AND GLOBULAR CLUSTERS. 47

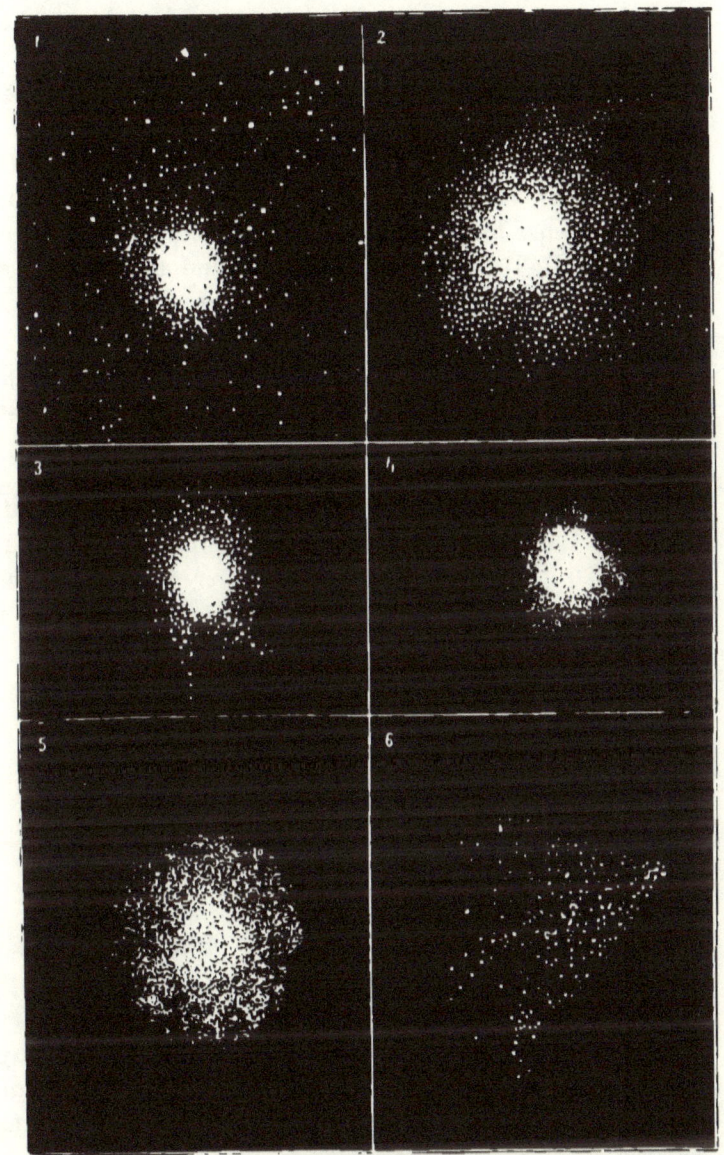

STAR-CLUSTERS.

1. In Libra.   2. In Hercules.   3. In Capricornus.   4. In Serpens.
5. In Aquarius.   6. In Gemini.

out of our universe, and which we must regard as other universes, are at such immeasurable distances, and are therefore so faint, that even the most powerful telescopes fail to reveal their real shape and boundaries. There is a gradual fading away at the edge, the last traces of which appear either as a luminous mist or cloud-like filament, which becomes finer till it ceases altogether to be seen. The Dumb-Bell Cluster, in Vulpecula, and the Crab Cluster, in Taurus, both of which have been resolved into stars, are instances of this.

It is proper to say, however, that some astronomers believe all the visible star-clusters and nebulæ to belong to our star-system or universe, which, if this be so, must include within itself miniatures of itself on a greatly reduced scale.

90. In some of these star-clusters, the increase of brightness from the edge to the centre is so rapid as to make it appear that the stars are actually nearer together at the centre than they are at the edge; in fact, that there is a real condensation toward the centre.

CHAPTER II.

NEBULÆ.

91. **Nebulæ under the Telescope.**—The term *nebula* was formerly applied to every thing in the sky which appeared cloud-like to the naked eye or in a telescope. Every time, however, a new telescope more powerful than any before used was brought to bear on them, numbers of what were till then called nebulæ, and about which as

---

Clusters. 89. Describe some of the very distant clusters. Mention two of these.
90. What would seem to follow from the increase of brightness toward the centre of some of the clusters?
91. To what was the term *nebula* formerly applied? What revelations were

nebulæ nothing was known, were found to be star-clusters, some of them of very remarkable forms, so distant that the smaller telescopes, powerful though they were, had failed to resolve them into distinct stars. Now, this is what has happened ever since the discovery of telescopes. Hence it was thought by some that all the so-called nebulæ were, in reality, nothing but distant star-clusters.

92. One of the most important discoveries of modern times, however, has furnished evidence of a fact long ago conjectured by some astronomers—namely, that some of the nebulæ are something different from masses of stars, and that their cloud-like appearance is due to something else besides their distance and the insufficient power of our telescopes. This discovery is so recent that there has not yet been time to sort out the real from the apparent nebulæ. We are obliged, therefore, still to accept as nebulæ all formerly classed as such which up to this time have not been resolved into stars.

Fig. 21.—Great Nebula of Orion.

93. **Classification.**—Nebulæ may be divided into five classes:—1. Irregular Nebulæ. 2. Ring and Elliptical Nebulæ. 3. Spiral or Whirlpool Nebulæ. 4. Planetary Nebulæ. 5. Nebulæ surrounding stars.

made, as more powerful telescopes were used? What inference was drawn from this? 92. What has since been discovered respecting some of the nebulæ? 93. Into how many classes may nebulæ be divided? Name them. 94. To which

94. Irregular Nebulæ.—Some of the irregular nebulæ are visible to the naked eye on a dark night. Among these is the great nebula of Orion (Fig. 21), in the part of the constellation occupied by the sword-handle and surrounding the multiple star *Theta* (θ). The nebulosity near the stars has the appearance of separate flakes, and is of a greenish-white tinge. There seems no doubt that the shape of this nebula and the position of its brightest portions are changing. One part of it appears, in a powerful telescope, startlingly like the head of a fish. On this account it has been termed the Fish-mouth Nebula.

Two fine irregular nebulæ are visible in the southern hemisphere: one is in the constellation Dorado, the other surrounds *Eta* (η) *Argûs*. The latter occupies a space equal to about five times the apparent area of the Moon.

95. Ring and Elliptical Nebulæ.—We have classed the ring and elliptical nebulæ together, because probably the latter are ring-nebulæ looked at sideways. The finest ring-nebula is in the constellation Lyra, not far from the star Vega. As seen by Sir John Herschel, it presented the appearance of an oval ring surrounding a darker space (see Fig. 22, No. 1), the uniform pale glimmer of which resembled a light gauze stretched across the ring. Lord Rosse's more powerful telescope has since partially resolved the

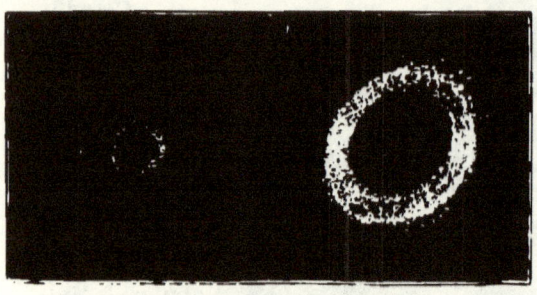

Fig. 22.—Ring-Nebula in Lyra.

of these classes does the great nebula of Orion belong? Describe this nebula. What name has been given to it, and why? What irregular nebulæ are visible in the southern hemisphere? 95. Why are the Ring and Elliptical Nebulæ classed together? Which is the finest ring-nebula? Describe it, as seen by Sir John Herschel. As seen through Lord Rosse's telescope. Where is there a fine

ring into luminous points (see Fig. 22, No. 2), and has shown parallel lines in the opening and a fringe of light about the outside border.

Near the beautiful triple star *Gamma* (γ) *Andromedæ* is a fine specimen of an elliptical nebula, having two stars near the extremities of the major axis of the ellipse.

Fig. 23.—ELLIPTICAL NEBULA near γ *Andromedæ*.

96. Spiral Nebulæ.—The spiral or whirlpool nebulæ are represented by that in the constellation Canes Venatici. In an ordinary telescope it presents the appearance of two globular clusters, one of them surrounded by a ring at a considerable distance, the ring varying in brightness, and being divided into two in a part of its length. But in a larger instrument the appearance is entirely changed. The ring turns into a spiral coil of nebulous matter, and the outlying mass is seen connected with the main mass by a curved band.

Fig. 24.—SPIRAL NEBULA IN CANES VENATICI.

In the constellations Pisces and Virgo we have other examples of this strange phenomenon (the 33d and 99th

---

specimen of an elliptical nebula? 96. In what constellation is there a remarkable spiral nebula? Describe it. In what other constellations do spiral nebulæ

in Messier's catalogue), which indicate the action of stupendous forces of a kind unknown in our own universe.

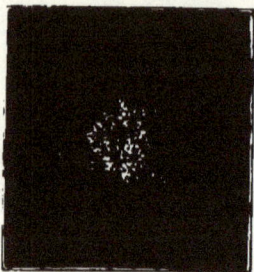

FIG. 25.—PLANETARY NEBULA IN URSA MAJOR.

97. **Planetary Nebulæ.**—These were so called by Sir John Herschel. They are circular or slightly elliptical in form, and shine with a planetary and often bluish light. One in Ursa Major will serve as a specimen.

98. **Nebulæ surrounding Stars.**—We come lastly to the nebulæ surrounding stars, or nebulous stars. The stars thus surrounded are apparently like all other stars, save in the fact of the presence of the appendage; nor does the nebula give any signs of being resolvable with our present telescopes. *Iota* (ι) *Orionis, Epsilon* (ε) *Orionis*, 8 *Canum Venaticorum*, and 79 *Ursæ Majoris*, belong to this class.

FIG. 26.—NEBULOUS STAR, ι Orionis.

99. **Brightness of the Nebulæ.**—Like the stars, the nebulæ differ in brightness, but as yet they have not been divided into magnitudes. This, however, has been done in a manner by determining the space-penetrating or light-grasping power of the telescopes powerful enough to render them visible.

Thus, it has been estimated that Lord Rosse's great Reflector, the most powerful instrument as yet used in such inquiries, penetrates 500 times farther into space than the naked eye can; hence a nebula which this telescope just renders visible must be 500 times farther off than a star of the sixth magnitude. Now, as light requires 120 years to reach us from such a star, the tele-

---

occur? 97. From whom did the planetary nebulæ receive their name, and why? What is their form? Where does one occur? 98. What is the last class of nebulæ? Describe the nebulous stars. Mention four of this class. 99. How do the nebulæ compare with each other in brightness? How has their magnitude in a manner been determined? Illustrate this in the case of Lord Rosse's tele-

scope referred to penetrates so profoundly into space that no star can escape its scrutiny, unless at a distance that it would take light sixty thousand years to traverse.

An idea of the extreme faintness of the more distant nebulæ may be gathered from the fact, that the light of some of those visible in an instrument of moderate size has been estimated to range from $\frac{1}{1500}$ to $\frac{1}{20000}$ of the light of a sperm-candle consuming 158 grains of material per hour, viewed at the distance of a quarter of a mile; that is, *such a candle a quarter of a mile off is from 1,500 to 20,000 times more brilliant than these nebulæ.*

100. **Variable Nebulæ.**—The phenomena of variable, lost, new, and temporary stars, have their equivalents in the case of the nebulæ, the light of which, it has been lately discovered, is in some cases subject to great variations.

In 1861 it was found that a small nebula, discovered in 1856 in Taurus, near a star of the tenth magnitude, had disappeared, the star also becoming dimmer. In the next year the nebula regained its brightness. Another nebula, which in May, 1860, appeared as a star of the seventh magnitude, during the next month recovered its nebulous appearance.

101. **Distribution of the Nebulæ.**— In Art. 48 the marked character of the distribution of the stars of our universe, giving rise to the appearance of the Milky Way, was pointed out. The distribution of the nebulæ, however, is very different; in general, they lie out of the Milky Way, so that they are either less condensed there, or the *visible* universe (as distinguished from our own *stellar* universe) is less extended in that direction. They are most numerous in a zone which crosses the Milky Way at right angles, the constellation Virgo being so rich in

---

scope. How does the light of some nebulæ visible through a telescope of moderate size compare with that of a candle? 100. Give examples of variable nebulæ. 101. How are the nebulæ distributed? Where are they most numerous?

them that a portion of it is termed the nebulous region of Virgo. In fact, not only is the Milky Way the poorest in nebulæ, but the parts of the heavens farthest from it are the richest.

102. **Physical Constitution of the Nebulæ.**—We now come to the question, What is a nebula? The answer is—*A true nebula is a mass of incandescent or glowing gas*, and there are indications that the gases in question are nitrogen and hydrogen. This fact, the fruit of the brilliant discovery before alluded to (Art. 92), forever sets at rest the question so long debated, as to the existence of a Nebulous Fluid in space.

When, therefore, we see closely-associated points of light in a nebula, we must not suppose that the latter is necessarily resolvable into stars. These luminous points, in some nebulæ at least, must be looked upon as themselves gaseous bodies, denser portions probably of the great nebulous mass. It has been suggested that the apparent permanence of general form in a nebula is kept up by the continual motions of these denser portions.

103. **The Nebular Hypothesis,** given to the world before the existence of a nebulous fluid was proved, supposes that there once existed in space a great, chaotic, nebulous mass, endowed with a kind of whirlpool motion, which, gradually condensing through the mutual attraction of its particles, formed the countless suns distributed through space; that the planets were formed by the condensation of rings of matter successively thrown off by the central mass, and the satellites by the condensation of matter thrown off in like manner by their primaries. It may take years to prove, or disprove, this hypothesis; but the tendency of recent observations is to show its correctness.

---

102. Of what is a nebula composed? When we see closely-associated points of light in a nebula, what must we not suppose? What may these luminous points be? 103. What is the substance of the Nebular Hypothesis? What is the bearing of recent observations?

CHAPTER III.

THE SUN.

104. **The Sun.**—We shall now consider the star nearest to us, which dazzles the whole family of planets by its brightness, supports their inhabitants by its heat, and keeps them in bounds by its weight. In almanacs and astronomical treatises, the Sun is denoted by either of the following signs: ☉ or ⊙.

105. **The Sun's Disk.**—The Disk of a heavenly body is its face, as it appears projected on the sky. The Sun's disk is a perfect luminous circle. Hence, as we know that the Sun revolves on its axis (Art. 110), we conclude that its form is that of a perfect sphere.

The Sun's disk varies slightly in size, according to the Earth's distance from the Sun, being largest about January 1st, when we are nearest to it, and smallest about July 1st, when we are farthest off. If the mean size of the disk (presented to us about the 1st of April and October) be represented by 100, its greatest size will be 107, and its least 94.

106. **Relative Brilliancy and Size.**—The brilliancy of the Sun, compared with that of the other stars, is so great that it is difficult at first to look upon it as in any way related to those feeble twinklers. This difficulty, however, is soon dispelled when we consider that its distance from us is less than $\frac{1}{100000}$ of that of the nearest star, *Alpha* (a) *Centauri*. Removed as far as the latter is from us, our Sun would be a star of the second magnitude; and, removed to the mean distance of the first-magnitude stars,

---

104. What are we next to consider? By what signs is the Sun denoted?
105. What is meant by the Disk of a heavenly body? What is the Sun's disk? Hence, as we know that the Sun turns on its axis, what do we conclude respecting its shape? When is the Sun's disk largest, when smallest, and why? 106. How does the light of the Sun compare with that of the stars? How is this difference explained? How would the Sun look, if removed to the mean dis-

it would be just visible to the unaided sight as a star of the sixth magnitude.

Our Sun is, therefore, by no means one of the largest stars. If we assume that the light given out by Sirius, for instance, is no more brilliant than our sunshine, that star would be equal in bulk to more than 3,000 Suns.

107. **Distance and Diameter.**—The mean distance of the Sun from the Earth is now known to be about 91,000,000 miles. These figures, as in the case of the distances of the stars, fail to convey any definite idea to the mind. Were there a railroad from the Earth to the Sun, a train going night and day at the rate of 30 miles an hour, and starting on the 1st of January, 1870, would not reach the Sun till about the middle of the year 2208.

108. The Sun's distance being known, it is easy to determine its size. The distance from one side of the Sun to the other, through its centre—or, in other words, the *diameter* of the Sun,—is 852,584 miles. If the Sun were so placed that its centre coincided with that of the Earth, this immense luminary would not only fill the whole orbit of the Moon, but extend beyond it three-fourths of the Moon's distance from the Earth. A train going at the speed named above would accomplish the journey round our Earth in a little over a month; a railway journey round the Sun, the same speed being maintained, would require more than ten years.

If we represent the Sun by a globe about two feet in diameter, a pea at the distance of 430 feet will represent the Earth; and the nearest fixed star would be represented by a similar globe placed at the distance of 9,000 miles.

109. **Volume and Mass.**—More than 1,200,000 Earths would be required to make one Sun. Astronomers ex-

---

tance of the 1st-magnitude stars? How does the Sun compare in size with Sirius? 107. How far is the Sun from the Earth? Give some idea of this distance, by telling how long it would take to travel it by rail. 108. What is the length of the Sun's diameter? Give an idea of this distance. How may we represent the Sun, the Earth, and the nearest fixed star? 109. What is the dif

press this by saying that the *volume* of the Sun is over 1,200,000 times greater than that of the Earth. But as the matter of which the Sun is composed weighs only one-quarter as much, bulk for bulk, as that of the Earth, 300,000 Earths only would be required in one scale of a balance to weigh down the Sun in the other. That is, the *mass*, or weight of the Sun, is 300,000 times greater than that of our Earth.

110. **Rotation.**—The Sun, like the Earth or a top when spinning, turns round on an axis; this rotation was discovered by observing the spots on its surface, about which we shall presently have much to say. It is found that the spots always make their first appearance on the same side of the Sun; that they travel across it in from twelve and a half to fourteen days, and then disappear on the other side. This is not all: if they be observed in June, they go *straight* across the sun's disk with a dip downward; if in September, they cross in a curve; while in December they go straight across again, with a dip upward, and in March their paths are again curved, but this time in the opposite direction.

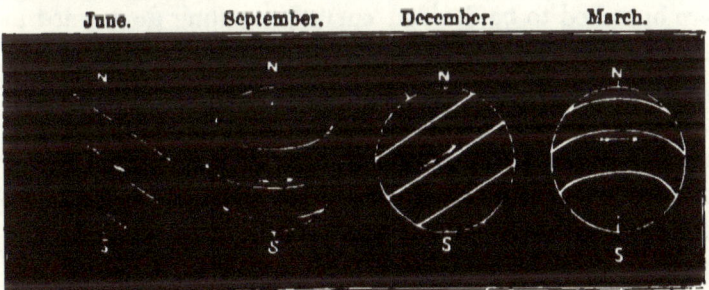

Fig. 27.—Apparent Paths of the Spots across the Sun's Disk, as seen from the Earth at different times of the year. The arrows show the direction in which the Sun rotates.

111. **The Plane of the Ecliptic.**—It is important that

---

ference between *volume* and *mass*? How does the Sun compare with the Earth in volume? How, in mass? 110. What has been found, by observing the spots on the Sun? What appearances do these spots present? 111. What is meant by

we make this perfectly clear. We know that the Earth goes round the Sun once a year. It has been found, also, that its path is level—that is to say, the Earth in its journey does not go up or down, but always straight on; we may imagine it as floating round the Sun on a boundless ocean, in which both Sun and Earth are half immersed. We shall see further on that this level—called the Plane of the Ecliptic—is used by astronomers in precisely the same way as we commonly use the sea-level. We say, for instance, that such a mountain is so high *above the level of the sea.* Astronomers say that such a star is so high *above the plane of the ecliptic.*

112. **Inclination of the Sun's Axis.**—We have imagined the Earth and Sun to be floating in an ocean up to the middle. Now, if the Sun were quite upright, the spots would always seem at the same distance above the level of our ocean. But this, we have found, is not the case. From the two opposite points of the Earth's path which it occupies in June and December, the spots are seen to describe straight lines across the disk, while midway between these points (in September and March) their paths are observed to be decided curves, rounding downward in the one case and upward in the other (see Fig. 27). A moment's thought will show that these appearances can arise only from an inclination in the Sun's axis. The Earth in its annual revolution attains in September a point at which the Sun's axis is inclined toward it; and in March reaches the opposite point of its orbit, at which the Sun's axis is inclined away from it.

113. **Time of Rotation.**—It has been found that the spots, besides having an apparent motion, caused by their being carried round by the Sun in its rotation, have a mo-

---

the Plane of the Ecliptic? What use is made of it by astronomers? 112. What is found to be the case, with regard to the Sun's axis? How is this inclination proved? Why do the paths of the spots curve in different directions in September and March? 113. What motion have the spots besides their *apparent* motion?

tion of their own. This *proper motion*, as distinguished from their *apparent motion*, has recently been thoroughly investigated, and accounts for the great difference in the periods which different observers have assigned to the Sun's rotation. As already stated, this rotation has been deduced from the time taken by the spots to cross the disk; but it now seems that all sun-spots have a motion of their own, the rapidity of which varies regularly with their distance from the solar equator—that is, the line half-way between the two poles of rotation. The spots near the equator travel faster than those away from it: so that, if we take an equatorial spot, we shall say that the Sun rotates in about twenty-five days; whereas, if we take one half-way between the equator and the poles, in either hemisphere, we shall say that it rotates in about twenty-eight days.

We are still, therefore, ignorant of the exact time of the Sun's rotation; for, if it is a solid mass, it can of course have but one period—and which of the two named above it may be, if either of them, we have no means of telling.

114. **Telescopic Appearance.**—We have now considered the distance and size of the Sun; we have found that, like our Earth, it rotates on its axis, and we have determined the direction in which the axis points. We must next try to learn something of the appearance it presents when viewed through a telescope, and of its nature or physical constitution. On this latter point our knowledge is not yet complete. This, however, is little to be wondered at. We have gleaned so many facts, at stupendous distances the very statement of which is almost meaningless to us, that we forget that our mighty Sun, in spite of its brilliant light and fostering heat, is still some 91,000,000 miles

---

For what does this proper motion account? How do the sun-spots differ, as regards their proper motion? What is the exact time of the Sun's rotation? 114. What keeps us from knowing more about the physical constitution of the Sun? What caution is given, with respect to looking at the Sun? 115. What are the first things that strike us, on looking at the Sun through a powerful

away; and that, even though we employ the finest telescope, we can only observe the various phenomena as we should do with the naked eye at a distance of 180,000 miles.

To look at the Sun through a telescope, without proper appliances, is a very dangerous affair. Several astronomers have lost their eyesight by so doing, and the student should not use even the smallest telescope without proper guidance.

115. **Sun-spots.**—The first things which strike us on the Sun's surface, when we look at it with a powerful telescope, are dark spots. On the opposite page we give drawings of a very fine one, visible on the Sun in 1865. The spots are not scattered all over the Sun's disk, but are generally limited to those parts of it a little above and below the Sun's equator, which is represented by the middle lines in Fig. 27.

116. The spots float, as it were, in what, as we have already seen in the case of the stars, is called the photosphere; the half-shade shown in the spot is called the **penumbra**; inside the penumbra is a still darker shade, called the **umbra**, and inside this again is the **nucleus**. Figs. 3 and 4 on the opposite page will render this perfectly clear. The white surface is the photosphere; the half-tones represent the penumbra; the dark, irregular central portions, the umbra; and the blackest parts in the centre of these dark portions, the nucleus.

117. Sun-spots are cavities, or hollows, in the photosphere, and these different shades represent different depths.

118. Diligent observation of the umbra and penumbra, with powerful instruments, reveals to us the fact that change is incessantly going on in the region of the spots. Sometimes changes are noticed even within an hour: here

---

telescope? How are these spots situated? 116. What is the Photosphere? The Penumbra? The Umbra? The Nucleus? 117. What are Sun-spots? 118. What is constantly going on in the region of the spots? Describe some of these

SUN-SPOTS.

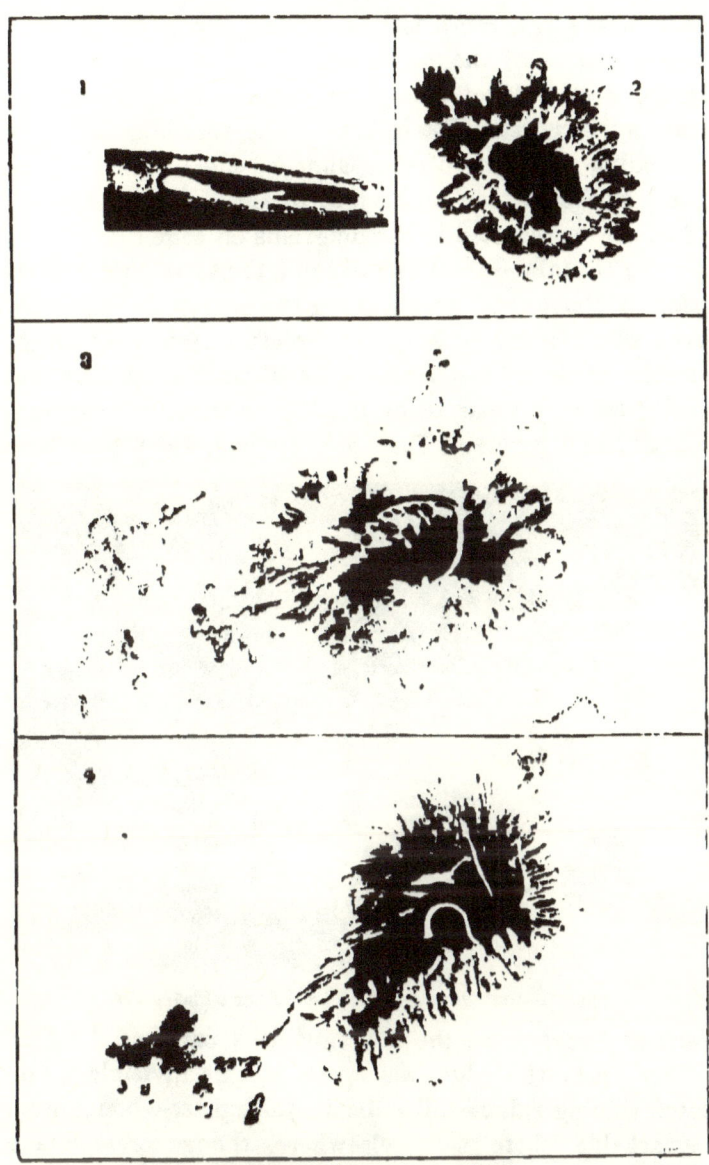

THE GREAT SUN-SPOT OF 1865.
1. The spot entering the Sun's disk, Oct. 7th (foreshortened view). 2. Its appearance, Oct. 10th. 3. Central view, Oct. 14th, showing the formation of a bridge, and the nucleus. 4. Its appearance, Oct. 16th.

part of the penumbra is seen sailing across the umbra; here a portion of the umbra is melting from sight; here, again, is an evident change of position and direction in masses which retain their form. The enormous changes, extending over tens of thousands of square miles of the Sun's surface, which took place in the great sun-spot of 1865, are represented in the diagrams on page 61.

119. **Faculæ.**—Near the edge of the solar disk, and especially about spots approaching the edge, it is quite easy, even with a small telescope, to discern certain very bright streaks of diversified form, quite distinct in outline, and either entirely separate or uniting in various ways into ridges and net-work. These appearances, which have been

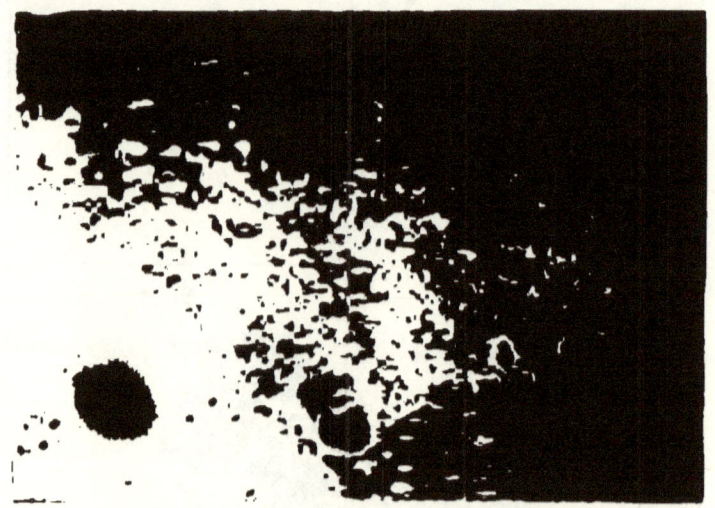

FIG. 28.—SUN-SPOTS AND FACULÆ. From a Photograph.

termed *Faculæ*, are the most brilliant parts of the Sun. Where, near the edge, the spots become invisible, undulated shining ridges still indicate their place—being more remarkable there than elsewhere, though everywhere traceable in good observing weather. Faculæ may be

changes. 119. What are Faculæ? What is said of their size? How do they sometimes lie, as regards spots? 120. How does the Sun's surface look, where

APPEARANCES ON THE SUN'S DISK.

of all magnitudes, from hardly-visible, softly-gleaming, narrow tracts 1,000 miles long, to continuous complicated ridges 40,000 miles and more in length, and from 1,000 to 4,000 miles broad. Ridges of this kind often surround a spot, and hence appear the more conspicuous; such a ridge is shown in Fig. 1, page 61. Sometimes there appears a very broad white platform round the spot, and from this white crumpled ridges pass in various directions.

120. **Other Appearances on the Sun's Disk.**—The whole surface of the Sun, except those portions occupied by the spots, is coarsely mottled; and, indeed, the mottled appearance requires no very great optical power to render it visible. Viewed through a large telescope, the surface seems to be made up principally of luminous masses, called by Sir William Herschel *corrugations*, and described by other observers as resembling "rice-grains," "granules," etc.

FIG. 20.—"WILLOW-LEAVES" IN A SUN-SPOT. *A*, tongue of facula stretching out into the umbra. *B*, clouds. *C*, layers of "willow-leaves" in the penumbra.

121. The term *willow-leaves* has been appropriately applied to appearances sometimes observed in the penumbræ of spots. They consist of elongated masses of unequal brightness,

---

It is not covered with spots? Of what does it seem to be made up, when viewed through a large telescope? 121. What is meant by *willow-leaves*? 122. What is

so arranged that for the most part they point like so many arrows to the centre of the nucleus, giving to the penumbra a radiated appearance. At other times, and occasionally in the same spot, the jagged edge of the penumbra projecting over the nucleus has caused the interior edge of the penumbra to be likened to coarse thatching with straw.

122. There are darker or shaded portions between the granules, often pretty thickly covered with dark dots, like stippling with a soft lead-pencil; these are what have been called *pores* by Sir John Herschel, and *punctulations* by his father. Some of these are almost black, and are like excessively small eruptive spots.

123. When the Sun is totally eclipsed,—that is, as will be explained by-and-by, when the Moon comes exactly between the Earth and the Sun,—other appearances are unfolded to us, which the extreme brightness of the Sun prevents our observing under ordinary circumstances. The Sun's atmosphere is then seen to contain red masses of fantastic shapes, some of them quite disconnected from the Sun; to these the names of *red-flames* and *prominences* have been given. Now, as these bodies appear much brighter than the surrounding atmosphere, we conclude that they are hotter than the latter, as a bright fire is hotter than a dim one.

124. **Explanation of the Appearances on the Sun's Disk.**—Let us see if we can account for the appearances which the Sun's disk presents, when viewed through a powerful telescope. As the spots break out and close up with great rapidity, as changes both on a large and a small scale are constantly taking place on the surface, we can only infer that the photosphere of the Sun, and there-

fore of the stars, is of a cloudy nature. But while *our* clouds are made up of particles of water, the clouds *on the Sun* must be composed of particles of various metals and other substances in a state of intense heat. The photosphere is surrounded by an atmosphere composed of the vapors of the bodies which are incandescent in the photosphere.

It seems, also, that not only is the visible surface of the Sun entirely of a cloudy nature, but that the atmosphere is a highly-absorptive one. Thus when the clouds are highest they appear brightest—*we see faculæ*—because they extend high into the atmosphere, and consequently there is less atmosphere to obscure our view. Spots may be due to the absorption of a greater thickness of atmosphere, as they are hollows in the cloudy surface; or the whole of the cloudy surface may be cleared off in those parts from a surface beneath, which emits less light than the clouds.

The more minute features—the granules—are most probably the dome-like tops of the smaller masses of cloud, bright for the same reason that the faculæ are bright, but in a less degree. The fact that these granules lengthen out as they approach a spot and descend the slope of the penumbra, may be accounted for by supposing them to be elongated by the current which draws them down into a spot, as the clouds in our own sky are lengthened out when they are drawn into a current.

125. **The Sun, a Variable Star.**—Some spots cover millions of square miles, and remain for months; others are visible only in powerful instruments, and are of very short duration. There is a great difference in the number of spots visible from time to time; indeed, there is a *minimum period*, when none are seen for weeks together,

---

solar atmosphere? Under what circumstances do we see faculæ? To what are spots due? What are the granules? How is their lengthening out as they approach a spot accounted for? 125. What is found to be the case, as regards

and a *maximum period*, when more are seen than at any other time. The interval between two maximum or two minimum periods is about eleven years.

Now, as we must get less light from the Sun when it is covered with spots than when it is free from them, we may look upon it as *a variable star, with a period of eleven years.*

It has recently been shown that this period is in some way connected with the action of the planets on the photosphere. It is also known that the magnetic needle has a period of the same length, its greatest oscillations occurring when there are most sun-spots. Auroræ, and the currents of electricity which traverse the Earth's surface, are affected by a similar period. There seems, therefore, to be some connection between these things and the solar spots, though what it is we do not know.

126. **Elements in the Sun.**—We have before seen (Art. 83) what substances exist in a state of incandescence in some of the stars. In the case of the Sun we are acquainted with a greater number. Here is the list:—

Sodium.	Zinc.	Gold, *probable.*
Iron.	Calcium.	Cobalt, *doubtful.*
Magnesium.	Chromium.	Strontium, *ditto.*
Barium.	Nickel.	Cadmium, *ditto.*
Copper.	Hydrogen, *probable.*	Potassium, *ditto.*

The atmosphere of the Sun, like that of the stars, consists of the vapors of these and of other—yet unknown—substances, and extends to a height exceeding 80,000 miles above the visible surface.

127. **Benign Influences of the Sun.**—Let us now inquire into some of the benign influences spread broadcast by the Sun. We all know that our Earth is lit up by its

---

the size and duration of the sun-spots? As regards the periods of their occurrence? What conclusion is drawn respecting the Sun? With what does the occurrence of solar spots seem to be connected? 126. Mention some of the elements known to exist in the Sun. Of what does the solar atmosphere consist?

beams, and that we are warmed by its heat; but this by no means exhausts its benefits, which we share in common with the other planets that gather round its hearth.

128. And first, as to its *light*. We have already compared its light with that which we receive from the stars, but that is merely its *relative* brightness; we want now to know its actual or *intrinsic* brightness. It is clear, at once, that no number of candles can rival this brightness; let us therefore compare it with one of the brightest lights that we know of—the calcium light. The calcium light proceeds from a ball of lime made intensely hot by a flame composed of a mixture of hydrogen and oxygen playing on it. It is so bright, that we cannot look on it any more than we can on the Sun; but if we place it in front of the Sun, and look at both through a dark glass, the calcium light, though so intensely bright, looks like a black spot. In fact, Sir John Herschel has found that the Sun gives out as much light as 146 calcium lights would do, if each ball of lime were as large as the Sun and gave out light from all parts of its surface.

129. Then, as to the Sun's *heat*. The heat thrown out from every square yard of the Sun's surface is greater than that which would be produced by burning six tons of coal on it each hour. Now, we may take the surface of the Sun roughly at 2,284,000,000,000 square miles, and there are 3,097,600 square yards in each square mile. How many tons of coal must be burnt, therefore, in an hour, to represent the Sun's heat?

130. But the Sun sends out, or *radiates*, its light and heat in all directions; it is clear, therefore, that as our Earth is so small compared with the Sun, and is so far away from it, the light and heat the Earth can intercept is but a very small portion of the whole amount; in fact,

---

128. How does the brightness of the Sun compare with that of a calcium light?
129. Give some idea of the Sun's heat. 130. How much of this does the Earth get? How much do all the planets together receive? What would be the effect

we only grasp the $\frac{1}{1,500,000,000}$ part of it. All the planets together receive but one 227-millionth part of the solar light and heat.

The *whole* heat of the Sun collected on a mass of ice as large as the Earth would be sufficient to melt it in two minutes, to boil the water thus produced in two minutes more, and to turn it all into steam in a quarter of an hour from the time it was first applied.

131. But this is not all. There is something else besides light and heat in the Sun's rays, and to this something we owe the fact that the Earth is clad with verdure; that in the tropics, where the Sun shines always in its might, vegetable life is most luxuriant, and that with us the spring-time, when the Sun regains its power, is marked by a new birth of flowers. There comes from the Sun, besides its light and heat, **chemical force**, which separates carbon from oxygen, and turns the gas which, were it to accumulate, would kill all men and animals, into the life of plants. Thus, then, does the Sun build up the vegetable world.

132. Let us go a step farther. The enormous engines which do the heavy work of the world,—the locomotives which take us so smoothly and rapidly across a whole continent,—the mail-packets which bear us so safely over the broad ocean,—owe all their power to steam, and steam is produced by heating water by coal. We all know that coal is the product of an ancient vegetation; and vegetation is the direct effect of the Sun's action. Hence, without the Sun's action in former times we should have had no coal. The heavy work of the world, therefore, is indirectly done by the Sun.

133. Now for the light work. Let us take man. To work, a man must eat. Does he eat beef? On what was

---

of the whole heat of the Sun, collected on a mass of ice as large as the Earth?
131. What else, besides light and heat, do we owe to the Sun? What is the effect of this chemical force? 132. Show how the heavy work of the world is

the animal which supplied the beef fed? On grass. Does he eat bread? Of what is bread made? Of the flour of wheat and other grains. In these, and in all cases, we come back to vegetation, which is, as we have already seen, the direct effect of the Sun's action. Here again, then, we must confess that to the Sun is due man's power of work. In fact, all the world's work, with one trifling exception (tide-work, of which more hereafter), is done by the Sun; and man himself, prince or peasant, is but a little engine, which merely *directs* the energy supplied by the Sun.

134. **Is the Sun inhabited?**—This is a question more easily asked than answered. If the whole body of the Sun is an incandescent globe, of course no organized beings of whom we can conceive can live upon it. But if the incandescence is confined to its photosphere, as many think, and the surface of the globe itself is protected from its outer envelope by a dense atmosphere, which absorbs its intense light and is at the same time a non-conductor of heat, there is nothing to prevent it from being inhabited.

135. **The Future of the Sun.**—Will the Sun keep up forever a supply of the force that has been described? It cannot, if it be not replenished, any more than a fire can be kept in unless we put on fuel; any more than a man can work without food. At present, philosophers know not by what means it is replenished. As, probably, there was a time when the Sun existed as matter diffused through infinite space, the condensation of which matter has stored up its heat, so, probably, there will come a time when the Sun, with all its planets welded into its mass, will roll, a cold, black ball, through infinite space.

We have no evidence, however, of any loss of heat, even from century to century; and, if there is a loss,

---

done by the Sun. 133. Show how man's power of work is due to the Sun. 134. Is the Sun inhabited? 135. What is the probable future of the Sun? 136.

there will doubtless be sufficient heat left to supply the planets with all they need for thousands of years to come.

136. Such, then, is our Sun—the nearest star. Although some of the stars do not contain those elements which on the Earth are most abundant (*a Orionis* and *β Pegasi*, for instance, are worlds without hydrogen), still we see that, on the whole, the stars differ from each other, and from our Sun, only in special modifications, and not in general structure. There is, therefore, a probability that they fulfil an analogous purpose; and are, like our Sun, surrounded with planets, which they uphold by their attraction, and illuminate and energize by their radiation. Hence the probable past and future of the Sun are the probable past and future of every star in the firmament of heaven.

CHAPTER IV.

THE SOLAR SYSTEM.

137. **General Description.**—From the Sun we now pass to the system of bodies which revolve round it; and here, as elsewhere in the heavens, we come upon the greatest variety. We find *planets*—of which the Earth is one—differing greatly in size, and situated at various distances from the Sun. We find again a ring of little planets clustering in one part of the system; these are called *asteroids*, or *minor planets:* and we already know of at least two masses or rings of smaller planets still, some of them so small that they weigh but a few grains. These give rise to the appearances called *meteors, bol'i-des,* or

---

Reasoning by analogy from the Sun, what may we suppose with respect to the stars?
137. What different bodies do we find in the Solar System? 138. How many

shooting-stars. We find also *comets*, some of which break in upon us from all parts of space, and then, passing round our Sun, rush back again; while others are so little erratic that they may be looked upon as members of the solar household. Besides these, there is another ring visible to us, under the name of the *zodiacal light*.

138. In the Solar System, then, we have **Eight large Planets**, named as follows, in the order of their distance from the Sun, and denoted in Almanacs, etc., by the signs appended to them respectively:—

1. Mercury, ☿   5. Jupiter, ♃
2. Venus, ♀   6. Saturn, ♄
3. Earth, ⊕   7. Uranus, ♅
4. Mars, ♂   8. Neptune, ♆

Two hundred and nineteen small Planets revolving round the Sun between the orbits of Mars and Jupiter. Their names are given in the Appendix, and they are denoted by numbers indicating the order of their discovery.

Meteoric Bodies, which at times approach the Earth's orbit, and occasionally reach the Earth's surface.

Comets.

The Zodiacal Light, a ring of apparently nebulous matter, the exact nature and position of which in the system are not yet determined.

139. **Explanation of the Signs.**—An explanation of the signs by which the eight large planets are denoted, may enable the student to remember them more easily.

Mercury was the messenger of the gods; the sign of the planet so called (☿) is deduced from the outline of his *caduceus*, or rod, which was entwined by two serpents and surmounted by a pair of wings. Venus, the goddess of beauty, has for her sign a circular looking-glass with a

---

large planets are there? Name them in the order of their distances from the Sun, and make the characters by which they are represented. How many asteroids are there? How are their orbits situated? Describe the Zodiacal Light. 139. Explain the meaning of the signs by which the eight large planets

handle (♀). The Earth's sign is a circle, denoting its shape (⊕). Mars, the god of war, has a round shield surmounted by a spear-head (♂). Jupiter's sign (♃) is derived from a capital *zeta* (Z), the initial of his Greek name, *Zeus*. Saturn, the god of time, is represented by the scythe with which he mows down the human race (♄). Uranus is denoted by a planet suspended from the cross-bar of an H, the initial of Herschel, its discoverer (♅). Neptune is known by his trident (♆).

140. **Historical Details.**—Of the eight large planets, Mercury, Venus, Mars, Jupiter, and Saturn, being visible to the naked eye, were known to the ancients. Uranus was discovered in 1781 by Sir William Herschel, from whom it was first commonly called Herschel. Its discoverer gave it the name of Georgium Sidus, in honor of King George III. Both these names, however, were discarded for the mythological one by which it is at present known.

141. Neptune was first seen and recognized as a planet by Dr. Galle, of Berlin, in 1846. The honor of its discovery is due to the French astronomer Le Verrier and the English Professor Adams.

The discovery of Neptune is one of the most astonishing facts in the history of Astronomy. As we shall see in the sequel, every body in our system affects the motions of every other body; and, after Uranus had been discovered some time, it was found that, on taking all the known causes into account, there was still something affecting its motion; it was suggested that this something was another planet, more distant from the Sun than Uranus itself. The question was, where was this planet, if it existed.

Adams and Le Verrier applied themselves, independently, to the solution of this problem, and arrived at results which showed a remarkable agreement, the positions

---

are distinguished. 140. Which of the planets were known to the ancients? When and by whom was Uranus discovered? What other names has it had? 141. To whom is the honor of the discovery of Neptune due? State the interest-

assigned the unknown planet respectively by the two astronomers not being a degree apart. Search was made in July, 1846, with the large telescope of the Cambridge Observatory, in the region indicated by the calculations of Mr. Adams; but no planet was recognized. In the following September, Le Verrier wrote to the Berlin observers, acquainting them with the results of his investigations, and requesting them to explore a certain part of the heavens where he imagined the planet then to be. Thanks to their superior star-map (which had not yet been published), the planet was discovered, in accordance with these instructions, that same evening.

142. The first of the asteroids, Ceres, was discovered in 1801 by the Sicilian astronomer Piazzi. Pallas was added to the list in 1802; Juno, in 1804; Vesta, in 1807; the rest have been discovered since 1844.

143. **A Suspected Planet.**—Besides the eight principal planets mentioned above, a ninth—quite small—is suspected to exist, between Mercury and the Sun, only thirteen million miles from the latter, and performing its revolution in about $19\frac{3}{4}$ days, in an orbit inclined to the ecliptic at an angle of 12°. A French physician, named Lescarbault, claimed to have discovered it crossing the Sun's disk in 1859. The name of **Vulcan** was assigned to it.

Other observers have, at different times, seen spots of a planetary character rapidly cross the disk of the Sun, which may turn out to have been transits of Vulcan; but up to the present time we can only say that the existence of such a planet is suspected—it is not proved. Le Verrier and other astronomers consider it not improbable, by reason of a certain disturbance in the motion of Mercury, for which a planet so situated would account.

---

ing facts connected with the discovery of Neptune. 142. Which four of the asteroids were first discovered, and when? 143. What is said respecting a ninth planet, whose existence is suspected? What appearances that have been observed may have been transits of Vulcan? What seems to make the existence

74   THE SOLAR SYSTEM.

144. Motions and Orbits of the Planets.—Let us begin by getting some general notions of the planetary motions and orbits. In the first place, *all the planets travel round the Sun in the same direction;* and that direction, looking down upon the system from the northern side of it, is *from west to east,* or, in other words, in the opposite direction to that in which the hands of a clock move. Secondly, *the paths of all the planets, and of many of the comets, are elliptical,* but some are very much more elliptical than others.

145. Next let the student turn back to Art. 111, in which we attempted to give an idea of the plane of the ecliptic. Now, the larger planets keep very nearly to this level, which is represented in the following figure:—

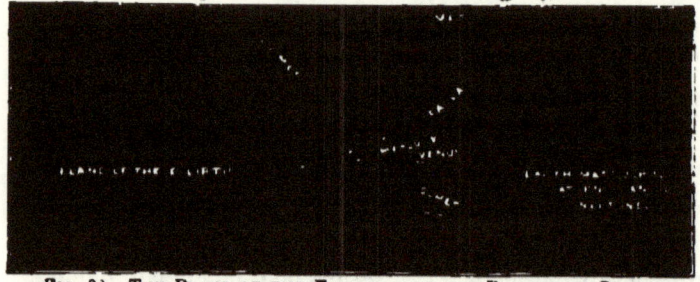

FIG. 30. - THE PLANE OF THE ECLIPTIC AND THE PLANETARY ORBITS.

The straight line we suppose to represent the Earth's orbit looked at edgeways. The other lines represent the orbits of some of the planets and comets seen edgeways in the same manner. The orbits of Mars, Jupiter, Saturn, Uranus, and Neptune, deviate so little from the plane of the ecliptic, that in our figure, the scale of which is very small, they may be supposed to lie in that plane. With some of the smaller planets and comets we see the case is very different. The latter, especially, plunge as it were

---

of such a planet probable? 144. What motion have all the planets? What is the shape of their orbits? 145. To what plane do the orbits of the larger planets keep very close? Which planet's orbit has the greatest dip? How are the

DISTANCES OF THE PLANETS. 75

down into the surface of our ideal sea, or plane of the ecliptic, in all directions, instead of floating on it or revolving in it.

146. **Moons.**—Again, as we thus find planets travelling round the Sun, so also do we find other bodies travelling round some of the planets. These are called Moons, or Satellites. The Earth has one Moon; Mars has two, Jupiter four, Saturn eight, Uranus four, and Neptune, according to our present knowledge, one.

147. **Motions of the Planets.**—All the planets revolve round the Sun and rotate on their axes in the same direction, *i. e.*, from west to east. The satellites also revolve round their primaries in the same direction, except those of Uranus and Neptune, which move from east to west.

148. **Distances from the Sun.**—Let us next inquire into the various distances of the planets from the Sun, bearing in mind that, as the orbits are elliptical, the planets are sometimes nearer to the Sun than at other times. The mean distances from the Sun, and the times of revolution, expressed in the Earth's days, are as follows:—

	Distance from the Sun in miles.	Period of revolution round the Sun. D.	H.	M.
Mercury,	35,393,000	87	23	15
Venus,	66,131,000	224	16	48
Earth,	91,430,000	365	6	9
Mars,	139,312,000	686	23	31
Jupiter,	475,693,000	4332	14	2
Saturn,	872,135,000	10759	5	16
Uranus,	1,753,851,000	30686	17	21
Neptune,	2,746,271,000	60126	17	20

149. The apparent size of an object varies with its dis-

---

orbits of the comets inclined, as regards the plane of the ecliptic? 146. What are Moons? What planets have moons, and how many has each? 147. What other motion besides that in their orbits have the planets? In what direction do the satellites revolve? 148. How far is the nearest planet from the Sun? How far is the farthest planet? What is the length of Mercury's year? Of Jupiter's?

tance; hence the solar disk must vary in size, as seen from the different planets, appearing largest to Mercury, which is nearest to it. Fig. 31 shows the relative size of the disk as seen from the several planets. It is well to remember that the relative size of the disk, as thus shown, represents also the relative amount of light and heat which the planets receive.

150. **Comparative Size of the Planets.**—The equatorial diameters of the planets are as follows:—

Diameter in Miles.		Diameter in Miles.	
Mercury,	2,962	Jupiter,	85,390
Venus,	7,510	Saturn,	71,904
Earth,	7,926	Uranus,	33,024
Mars,	4,920	Neptune,	36,620

151. We have before attempted to give an idea of the comparative size of the Earth and Sun, and of the distance between them; let us now complete the picture, with the aid of Sir John Herschel's familiar illustration. Taking a globe two feet in diameter to represent the Sun, Mercury would be a grain of mustard-seed, revolving in a circle 164 feet in diameter; Venus, a pea, in a circle 284 feet in diameter; the Earth, also a pea, at a distance of 430 feet; Mars, a rather large pin's head, in a circle of 654 feet; the asteroids, grains of sand, in orbits of from 1,000 to 1,200 feet; Jupiter, a moderate-sized orange, in a circle nearly half a mile across; Saturn, a small orange, in a circle of four-fifths of a mile; Uranus, a full-sized cherry, or small plum, in a circle more than a mile and a half across; and Neptune, a good-sized plum, in a circle about two miles and a half in diameter.

Fig. 32 will help to give an idea of the relative size of

Of Neptune's? 149. On what does the apparent size of an object depend? To which planet does the Sun look largest? To which, smallest? What does the relative size of the Sun's disk also represent? 150. Which planet has the greatest diameter? Which, the smallest? How does the Earth's diameter compare with that of Venus? With that of Jupiter? 151. Give Sir John Herschel's illustration of the comparative sizes and distances of the planets. What does Fig. 32

SIZE OF THE SUN'S DISK.

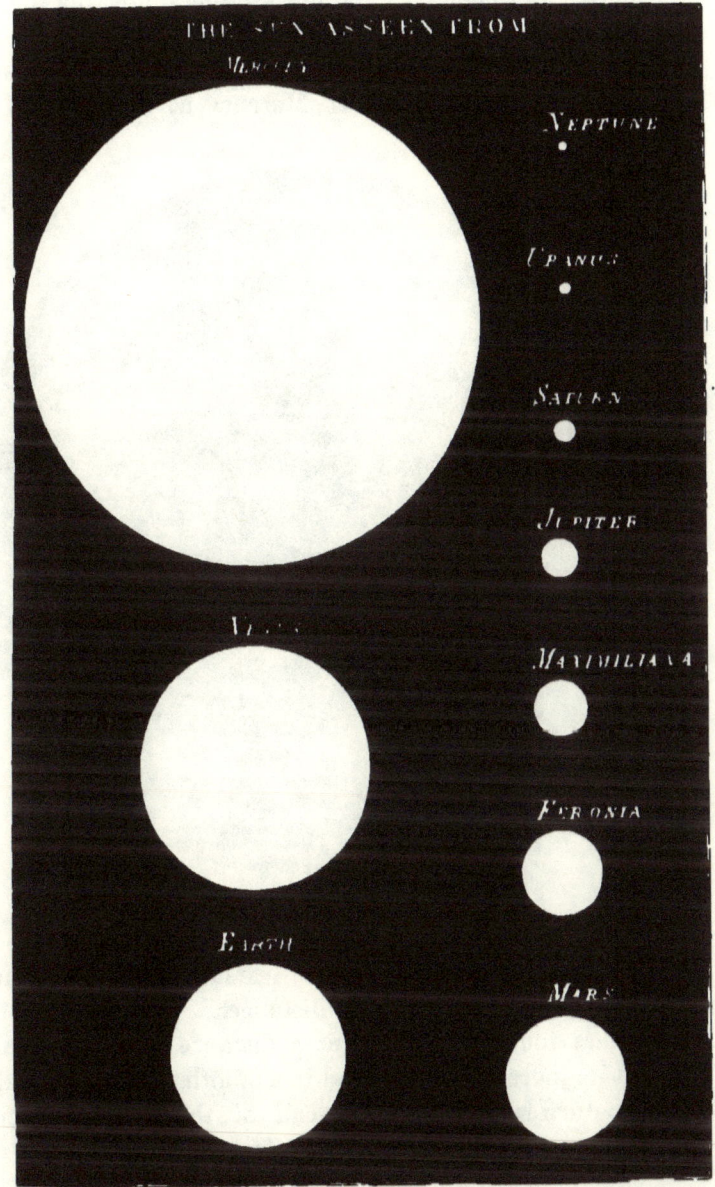

Fig. 31.—The Relative Size of the Sun, as seen from the Planets.

78 THE SOLAR SYSTEM.

the Sun and planets. The black circle represents the disk of the Sun. The disks of the several planets are represented by the white circles, on the same scale as that of the Sun, commencing with Mercury at the right of the upper line.

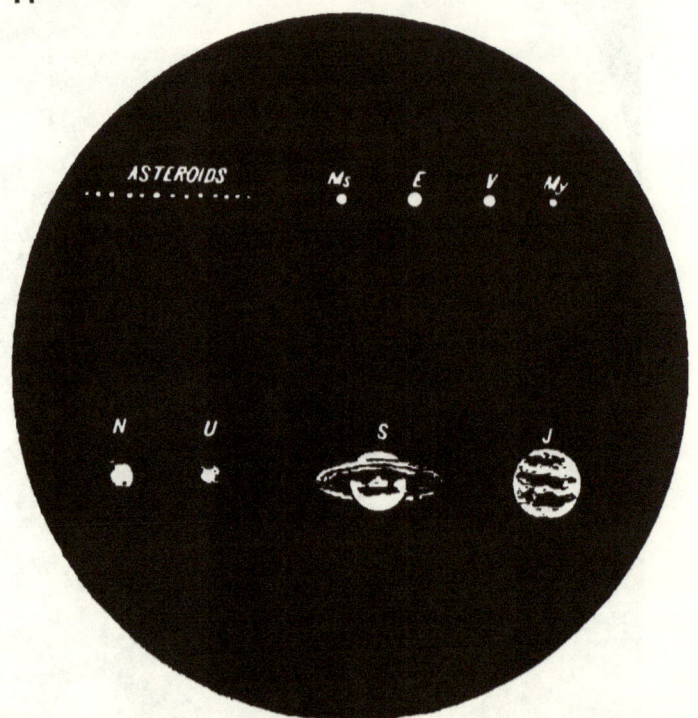

FIG. 32.—RELATIVE SIZE OF THE SUN AND PLANETS.

152. **Distances and Revolutions of the Satellites.**—The satellites revolve round their primaries, like the planets round the Sun, at different distances. Our solitary Moon courses round the Earth at a distance of 240,000 miles, and its journey is performed in a month. The first satellite of Saturn is only about one-half of this distance from its primary, and its journey is performed in less than a day.

The first satellite of Uranus is about equally near, and requires about two and a half days. The first satellite of Jupiter is about the same distance from that planet as our Moon is from us, and its revolution is accomplished in one and three-quarters of our days. The inner moon of Mars, only 6,000 miles from the centre of its primary, has a period of but 7 h. 38 m.—the shortest of all the satellites.

The diameter of the smallest planet—leaving the asteroids out of the question—is 2,962 miles. Among the satellites we have three bodies—the third and fourth satellites of Jupiter, and the sixth moon of Saturn—of greater dimensions than one of the large planets, Mercury, and nearly as large as another, Mars.

The distances and sizes of the planets and satellites are given in Tables II. and III. of the Appendix.

153. The *relative* distances of the planets from the Sun were known long before their *absolute* distances—just as we might know that one place was twice or three times as far away as another, without knowing the exact distance of either. When once the distance of the Earth from the Sun was known, astronomers could easily find the distance of all the rest from the Sun, and therefore from the Earth. Their sizes were next determined, for we need only to know the distance of a body and its apparent size, or the angle under which we see it, to determine its real dimensions.

154. **Volumes, Masses, and Densities of the Planets.**— In the case of a planet accompanied by satellites we can at once determine its weight, or *mass*, as will be shown hereafter; and when we have ascertained its weight, having already obtained its size or *volume*, we can compare the *density* of the materials of which it is composed

---

were known first—the *relative* distances of the planets from the Sun, or their *absolute* distances? When the distance of the Earth from the Sun was determined, what followed? What were next determined? 154. In what case can we at once determine the weight of a planet? If we know its size, what can we

with those we are familiar with here; having first also obtained experimentally the density of our own Earth.

155. Let us see what this word *density* means. To do this, let us compare platinum, the heaviest metal, with hydrogen, the lightest gas. The gas is, in round numbers, a quarter of a million times lighter than the metal, and therefore the same number of times less dense. If we had two planets of exactly the same size, one composed of platinum and the other of hydrogen, the latter would be a quarter of a million times less dense than the former. Now, if it seems absurd to talk of a hydrogen planet, we must remember that if the materials of which our system, including the Sun, is composed, once existed as a great nebulous mass extending far beyond the orbit of Neptune, as there is reason to believe, the mass must have been *more than* 200,000,000 *times less dense than hydrogen!*

156. Philosophers have found that the mean density of the Earth is a little more than five and a half times that of water; that is, our Earth is five and a half times heavier than it would be if it were made up of water. Looking at the planets together, we find that, as a general rule, they increase in density as we approach the Sun, Mercury being the densest, Venus and Mars agreeing very nearly with the Earth in density, Jupiter being only $\frac{1}{4}$ as dense as the Earth, and the more distant planets, Saturn, Uranus, and Neptune, being still less dense than Jupiter.

157. A table follows, showing the relative volume, mass, and density of the planets, the Earth's being represented by 100. The absolute volume of the Earth being, in round numbers, 259,400,000,000 cubic miles, and its weight 6,000,000,000,000,000,000,000 tons, the volume and weight of the other planets can be readily found from this table.

---

then do? 155. Illustrate the meaning of the word *density* by comparing platinum with hydrogen. 156. How does the density of the Earth compare with that of water? Comparing the other planets with the Earth as regards density, what do we find? 157. Which planet is about 300 times as heavy as the Earth? How

VOLUME, MASS, AND DENSITY

	Volume.	Mass.	Density.
Mercury,	5	7	124
Venus,	85	79	92
Earth,	100	100	100
Mars,	14	12	96
Jupiter,	138,743	30,000	22
Saturn,	74,689	9,000	12
Uranus,	7,236	1,300	18
Neptune,	9,866	1,700	17

158. **Summing up.**—To sum up, then, our first general survey of the Solar System, we find it composed of planets, satellites, comets, and several rings of meteoric bodies; the planets, both large and small, revolving round the Sun in the same direction, the satellites revolving round the planets. We have learned the mean distances of the planets from the Sun, and have compared the distances and times of revolution of some of the satellites. We have also seen that the volumes, masses, and densities of the planets have been determined. There is still much more to be learned, about both the system generally, and the planets particularly; but it will be best first to inquire somewhat minutely into the movements and structure of the Earth on which we dwell.

CHAPTER V.

THE EARTH.

159. We took the Sun as a specimen of the stars, because it was the nearest star to us, and we could therefore study it best; so now let us take our Earth, with which we should be familiar, as a specimen of the planets.

does Jupiter compare in density with the Earth? Which planet has the least density? 158. Sum up what we have thus far stated respecting the Solar System. 159. What body do we first consider, as a specimen of the planets? 160.

THE EARTH.

160. **Shape of the Earth.**—In the first place, the Earth is round. Had we no proof, we might have guessed this, because both Sun and Moon, and the planets observable in our telescopes, are round. But we have proof. The Moon, when eclipsed, enters the shadow of the Earth; and this shadow, as thrown on the bright Moon, is circular.

Moreover, if we watch ships putting out to sea, we lose first the hull, then the lower sails, until at last the highest

FIG. 33.—PROOF OF THE CURVATURE OF THE EARTH'S SURFACE.

parts of the masts disappear. So the sailor, when he sights land, first catches the tops of mountains, or other high objects, before he sees the beach or port. If the surface of the Earth were an extended plain, this would

---

What is the shape of the Earth? What proofs have we that the Earth is round?

THE SENSIBLE HORIZON.

not happen; we should see the nearest things and the largest things first. As it is, every point of the Earth's surface is the top, as it were, of a flattened dome interposed between us and distant objects. The inequalities of the land render this fact much less obvious on *terra firma* than on the surface of the sea.

Again, the roundness of the Earth has been proved by navigators, who, sailing in one direction, either east or west (as nearly as the different bodies of land would permit), have returned to the place from which they set out.

161. **The Sensible Horizon.**—On all sides of us we see a circle of land, or sea, or both, on which the sky seems to rest; this is called the Sensible Horizon. If we observe it from a little boat on the sea, or from a plain, this circle is small; but if we look out from the top of a ship's mast or from a hill, we find it greatly enlarged—in fact, the higher we go the more is the horizon extended, always however retaining its circular form. Now, the sphere is

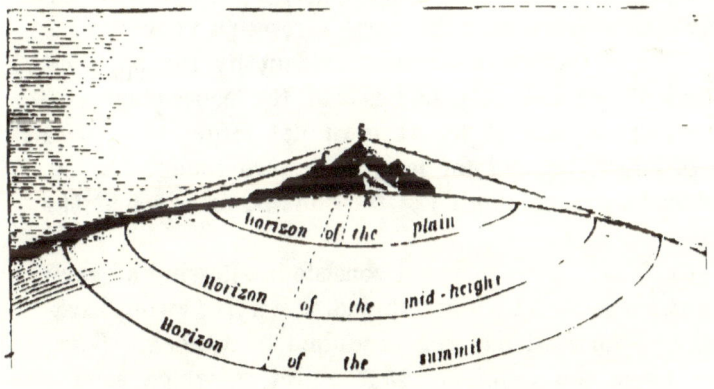

Fig. 34.—Horizons of the Same Place, at Different Heights.

the only figure which, looked at from any external point, is bounded by a circle; and as the horizons of all places are circular, the Earth is a sphere, or nearly so.

161. What is the Sensible Horizon? What proof of the Earth's roundness does

162. **Poles, etc.**—The Earth is not only round, but it rotates or turns on an axis, as a top does when it is spinning; and the names of *North Pole* and *South Pole* are given to those points where the axis would come to the surface if it were a great iron rod instead of a mathematical line. Half-way between these two poles, there is an imaginary line running round the Earth, called the *Equator* or *Equinoctial Line*.

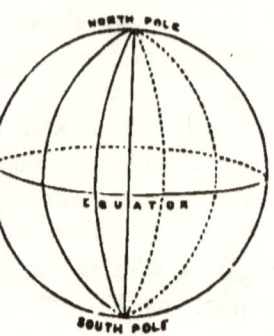

Fig. 35.—Poles and Equator.

The line through the Earth's centre from pole to pole, is called the *Polar Diameter;* the line through the Earth's centre from any point of the equator to the opposite point, is called the *Equatorial Diameter;* and one of these, as we shall see, is longer than the other.

163. **Proofs of the Earth's Rotation.**—We owe to the ingenuity of the French philosopher, Foucault, two experiments which render the Earth's rotation visible to the eye. For, although it is made evident by the apparent motion of the heavenly bodies and the consequent succession of day and night, we must not forget that these effects might be, and for long ages were thought to be, produced by a real motion of the Sun and stars round the Earth.

164. The first experiment consists in allowing a heavy weight, suspended by a fine thread or wire, to swing backward and forward like the pendulum of a clock. Now, if we move the beam or other object to which such a pendulum is suspended, we shall not alter the direction in which the pendulum swings, as it is easier for the thread

---

the sensible horizon afford? 162. What is meant by the North and the South Pole of the Earth? By the Equator? By the Polar Diameter? By the Equatorial Diameter? Which of these two diameters is the longer? 163. To whom are we indebted for having made the Earth's rotation on its axis visible to the eye? 164. Give an account of the first experiment. Where and how might such

which supports the weight to twist than for the heavy weight itself to alter its course when once in motion in any particular direction. Therefore, if the Earth were at rest, the swing of the pendulum would always be in the same direction with regard to the support and the surrounding objects.

Foucault's pendulum was suspended from the dome of the Pantheon in Paris, and a fine point at the bottom of the weight was made to leave a mark in sand at each swing. The marks successively made in the sand showed that the plane of oscillation varied with regard to the building. Here, then, was a proof that the building, and therefore the Earth, moved.

Such a pendulum swinging at either pole would make a complete revolution in 24 hours, and would serve the purpose of a clock were a dial placed below it with the hours marked. As the Earth rotates at the north pole from west to east, the dial would appear to a spectator, carried round like it by the Earth, to move under the pendulum from west to east, while at the south pole the Earth and dial would travel from east to west; midway between the poles, that is, at the equator, this effect, of course, is not noticed, as there the two motions in opposite directions meet.

165. The second experiment is based upon the fact that, when a body turns on a perfectly true and symmetrical axis, and is left to itself in such a manner that gravity is not brought into play, the axis maintains an invariable position; indeed, to maintain its position, it will even overcome slight obstacles. If, then, the axis of a heavy disk, so freely suspended that it is at almost entire liberty to turn in any direction, be made to point to a star, which is a thing outside the Earth, it will continue to point to it—

---

a pendulum be made to serve as a clock? How would the dial appear to move at the north pole? How, at the south pole? How, at the equator? 165. On what fact is the second experiment based? 166. What instrument does it em-

even turning, if the Earth's rotation makes it necessary, in order to keep the same absolute direction.

166. The Gyroscope is an instrument so made that a heavy disk, freely suspended and set in very rapid motion, shall be able to rotate for a long period, and that all disturbing influences, the action of gravity among them, may be as far as possible prevented.

Now, if the Earth were at rest, there would be no apparent change in the position of the axis, however long the wheel might continue to turn; but if the Earth moves and the axis remains at rest, there should be some difference. Experiment proves that there is a difference, and just such a difference as is accounted for by the Earth's rotation. In fact, if we so arrange the gyroscope that the axis of its rotation points to a star, it will remain at rest with regard to the star, while it varies with regard to surrounding objects on the Earth. This is proof positive that it is the Earth which rotates on its axis, and not the stars that revolve round it; for in the latter case the axis of the gyroscope would remain invariable with regard to the Earth, and change its direction with regard to the star.

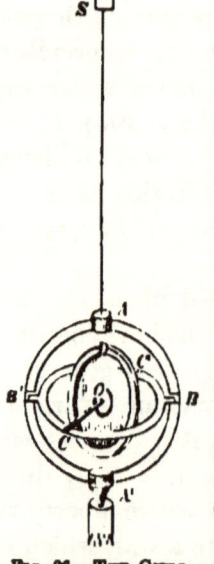

Fig. 36.—The Gyroscope.

Fig. 36 represents the interesting instrument with which the experiment just referred to is made. D is a heavy symmetrical metallic disk, mounted on an axis which passes through O, the centre of the disk, and is perpendicular to its two sides. This axis terminates in pivots $C\ C'$, which fit into holes made at opposite extremities of the diameter of a circular ring $B\ B'$, which is furnished with two knife-edges (like those of a balance), and so arranged that $B\ B'$ is the diameter of the ring perpendicular to $C\ C'$. The knife-edges rest in holes made at opposite extremities of the horizontal diameter of a vertical circle $A A'$, which is suspended by a fine wire from the fixed point S. At A' is a pivot, which rests in a small hole. All the pivots are highly polished, so that friction may be avoided as much as possible; and the different parts are so adjusted that O is the common centre of the disk and the rings. The axis $C C'$ may be made to point in any direction by moving first the ring $A A'$, and then the ring $B B'$, into proper positions.

To perform the experiment, $B B'$ is removed from its supports, and, the disk having been made to revolve rapidly, is then restored to its place. Whatever star $C C'$ is directed toward, it continues to point to as long as the disk rotates, and thus, as stated above, changes its position relatively to objects on the Earth; unless, indeed, the star be the polar star, in which case no change of direction will be observed.

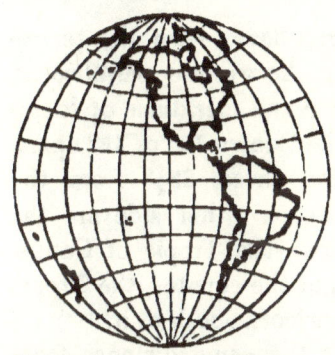

FIG. 37.—PARALLELS AND MERIDIANS.

167. **Imaginary Lines on the Earth's Surface.**—If we look at a terrestrial globe, we find that the equator is not the only line marked upon it. There are small circles parallel to the equator, called **Parallels**; and large circles, called **Meridians**, passing through both poles, and dividing the equator into equal parts. These lines are for the purpose of determining the exact position of a place upon the globe.

168. **Latitude.**—The distance of any place from the equator, measured in degrees (or 360ths) of its meridian, is called its Latitude. If north of the equator, it is said to be *in north latitude;* if south of the equator, *in south latitude.* As either pole is 90° distant from the equator, the greatest latitude a place can have is 90°.

169. **Longitude.**—But something else besides latitude is needed to define the position of a place. Accordingly, some meridian is taken,—in this country either the merid-

struction of the gyroscope. What is done to the instrument when the experiment is performed? 167. What circles do we find on a terrestrial globe? What is their object? 168. What is Latitude? What is the difference between North and South Latitude? What is the greatest latitude a place can have? 169. What else besides latitude is needed to define the position of a place? What is Longi-

ian of Washington, or that which passes through Greenwich, near London, where the principal observatory of England is situated; and the distance of the place from this First Meridian, as it is called, measured in degrees (or 360ths of its parallel), determines, with its latitude, its exact position. Distance from the first meridian, so measured, is called Longitude. Places east of the first meridian are said to be *in east longitude,* and those west of the first meridian *in west longitude.* As the distance half round the Earth is 180°, the greatest longitude a place can have is 180°.

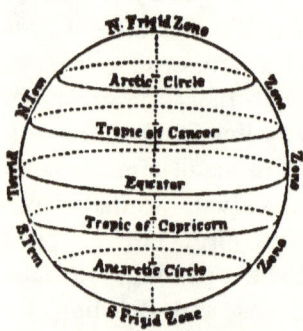

Fig. 38.—THE POLAR CIRCLES, TROPICS, AND ZONES.

170. **Zones.**—On the terrestrial globe we find parallels of latitude and meridians of longitude laid down 10° or 15° apart. Besides these, 23½° from the equator on either side are the **Tropics**,—the Tropic of Cancer north of the equator, the Tropic of Capricorn south of it.

At the same distance from the poles are the **Polar Circles**, the northern one being distinguished as the Arctic Circle, the southern as the Antarctic Circle. The tropics and polar circles divide the Earth's surface into five belts, or **Zones**—one *torrid,* two *temperate,* and two *frigid* zones, as shown in Fig. 38.

171. **Polar and Equatorial Diameter.**—The distance along the axis of rotation, from pole to pole, through the Earth's centre, is shorter than the distance through the Earth's centre from any point of the equator to the op-

---

tude? What meridian is generally taken as the First Meridian? What is the difference between East and West Longitude? What is the greatest longitude a place can have? 170. What are found 23½ degrees from the equator? What circles lie 23½ degrees from the poles? Into what do the tropics and polar circles divide the Earth's surface? How many degrees wide is each frigid zone? How wide is the torrid zone? How wide is each temperate zone? 171. What

THE EARTH'S DIAMETER.

posite one. In other words, the Polar Diameter (Art. 162) is shorter than the Equatorial Diameter. Their lengths are as follows:—

	Feet.	Miles.
Mean Equatorial Diameter,	41,848,380	7,925$\frac{434}{528}$
Polar Diameter,	41,708,710	7,899$\frac{121}{528}$

Difference in length, about 26$\frac{8}{11}$ miles.

This difference is but small; yet it proves that the Earth is not a sphere, but *an oblate spheroid* (see Art. 39).

172. The *mean* equatorial diameter is given above, for it is found that the equatorial circumference is not a perfect circle, but an ellipse, the difference between the major and minor axis of which is more than 1$\frac{2}{3}$ miles. The equatorial diameter which runs from longitude 14° 23'

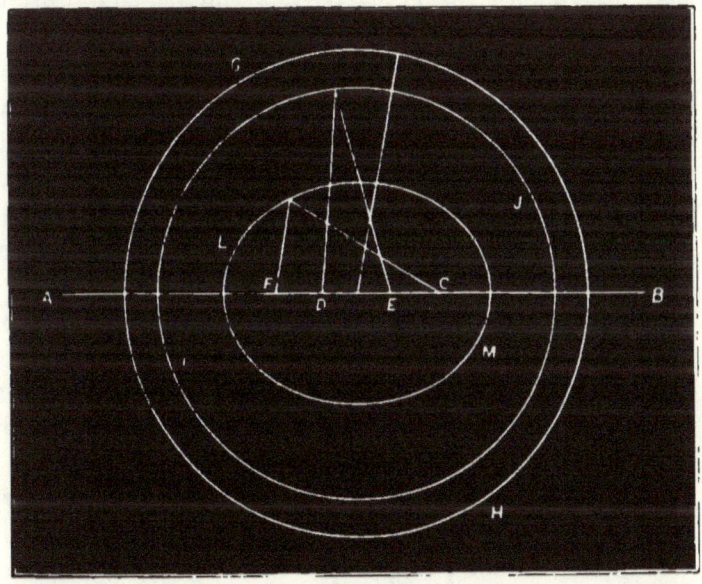

FIG. 59.—CIRCLE AND ELLIPSES. *G H*, a circle. *I J, L M*, ellipses of different eccentricities. *F, C*, foci of *L M*. *D, E*, foci of *I J*.

Is meant by the Polar Diameter of the Earth? The Equatorial Diameter? How do they compare in length? What, then, is the form of the Earth? 172. Why is the expression *mean* equatorial diameter used? 173. What produces the succession of day

east of Greenwich to 165° 37' west is over 1¼ miles longer than the one at right angles to it.

173. **Motions of the Earth.**—The Earth turns on its axis, or polar diameter, in 23h. 56m. In this time we get the succession of *day* and *night*, which is due to the Earth's rotation. The Earth also goes round the Sun, and the time in which that revolution is effected we call a *year*.

174. **Revolution round the Sun.**—The Earth and all the other planets move round the Sun in elliptical orbits. An ellipse was defined in Art. 35; its shape depends on the distance between its foci. It may be a decided oval, like LM in Fig. 39; or, if the foci are near together, it may have so little eccentricity as to be indistinguishable from a circle, as in the case of IJ, which looks as if it were parallel to the circle GH. The planetary orbits differ but little from circles.

175. **Perihelion and Aphelion.**—The Sun is not at the centre of the ellipses described by the planets, but at one of the foci. Hence every planet is nearer the Sun at one time of its revolution than at another. When nearest to the Sun, a planet is said to be *in perihelion* (from the Greek words περί, *near*, and ἥλιος, *the Sun*); when farthest off, *in aphelion* (ἀπό, *from*, and ἥλιος, *the Sun*).

176. The eccentricity of the ellipse described by the Earth is only $\frac{1}{60}$, so that when the orbit is represented on a small scale, as in Fig. 40, no deviation from a circle is perceptible. The Earth is 3,000,000 miles nearer the Sun in perihelion (at P, Fig. 40) than in aphelion (at A).

The Earth is in perihelion at present about January 1st, the time of the southern summer, and in aphelion about July 1st, the period of the northern summer. This

---

and night? What other motion has the Earth? 174. What is the shape of all the planetary orbits? How may ellipses differ? What kind of ellipses are the planetary orbits? 175. How is the Sun situated, as regards these ellipses? When is a planet said to be *in perihelion*, and when *in aphelion*? 176. What is the difference in the Earth's distance from the Sun at these two points? At what time of the year is the Earth in perihelion, and at what in aphelion?

greater nearness to the Sun intensifies the heat of the southern summer, and accounts for the fact that the temperature of this season is higher in Australia and Southern Africa than in corresponding latitudes north of the equator.

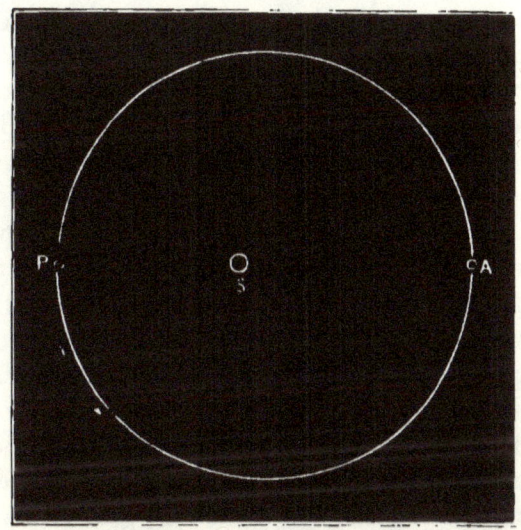

Fig. 40.—The Earth's Orbit. *S*, the Sun; *P*, the Earth in perihelion; *A*, the Earth in aphelion.

About 3,600 years before the creation of Adam, the Earth was nearest to the Sun during the summer of the northern hemisphere, and farthest off in the northern winter; which must have made the northern summer much hotter than it now is (according to Sir John Herschel, 23°), and the northern winter as much colder.

177. **Velocity of the Earth's Motions.**—Let us now inquire with what velocity the two motions of the Earth are performed.

As regards the diurnal motion, or rotation on the axis, it is clear that all the points on any meridian must make a complete revolution in the same time, while the circles or distances traversed in making such revolution diminish as we go from the equator to either pole. Hence, there is a material difference in the velocity of points in different latitudes. The poles have no rotary motion at all, and

---

.What is the consequence, as regards the southern summer? When must the northern summer have been hotter than it now is, and why? 177. Show why different parts of the Earth's surface have a different velocity of rotation. What

the regions about them very little. Points in the latitude of Paris have a velocity of about 330 yards a second; those in the latitude of Washington, about 375 yards; and those on the equator, about 507 yards. It has been demonstrated that the time of rotation has not varied one-hundredth of a second during the last two thousand years.

178. The velocity of the Earth in its orbit is constantly varying, being greatest when the Earth is in perihelion, as the Sun's attraction is then strongest. Its average rate is about 19 miles a second—more than a thousand times greater than that of the fastest locomotive. Two philosophers, who have attempted to determine the amount of heat that would be developed by the abrupt stoppage of the Earth in its orbit, tell us that it would suffice to melt the entire globe and reduce the greater part of it to vapor.

179. The reason why we are unconscious of moving with this immense rapidity is that we have never known any other condition, and that the whole bulk of the Earth and every object on and around it, including the atmosphere and clouds, participate in the motion.

180. **Inclination of the Earth's Axis.**—Now refer to Art. 112, in which we spoke of the position of the Sun's axis. We found that the Sun was not floating uprightly in our sea, the plane of the ecliptic; it was dipped down in a particular direction. So it is with our Earth. The Earth's axis is inclined in the same manner, but to a much greater extent (23° 27′ 24″). The direction of the inclination, as in the case of the Sun, is always the same.

181. **Effects of the Earth's Motions.**—We have, then, two completely distinct motions—one performed in a day, round the axis of rotation, which, so to speak, remains

---

Is the velocity in the latitude of Paris? In that of Washington? At the equator? What has been shown respecting the time of rotation? 178. What causes constant changes in the velocity of the Earth as it revolves round the Sun? What is the average rate of the Earth's motion in its orbit? 179. Why are we unconscious of this rapid motion? 180. State the facts respecting the inclination of the Earth's axis. 181. How many motions, then, has the Earth, and what do we

EFFECTS OF THE EARTH'S MOTIONS. 93

parallel to itself; the other round the Sun, performed in a year. To the former motion we owe *the succession of day and night;* to the latter, combined with the inclination of the Earth's axis, we owe *the seasons.*

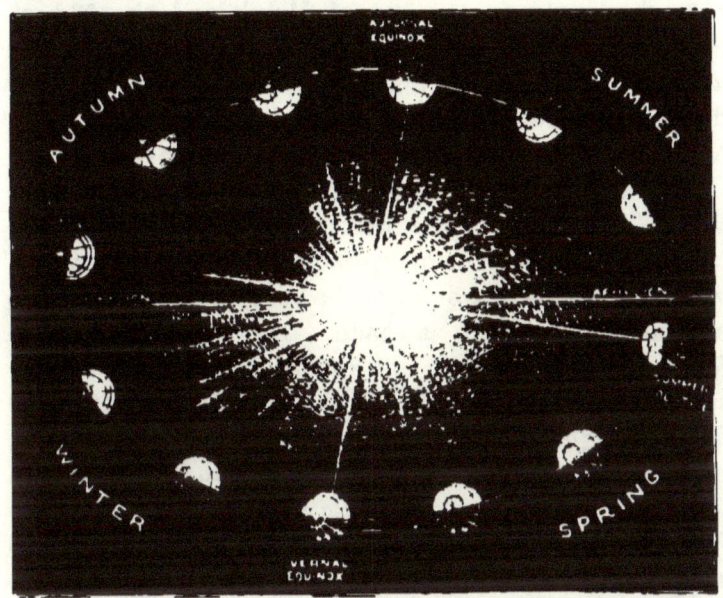

FIG. 41.—POSITION OF THE EARTH IN ITS ORBIT AT DIFFERENT SEASONS.

182. **Succession of Day and Night.**—Fig. 41 represents the orbit of the Earth, with the Sun at its centre. It also shows how the axis of the Earth is inclined, its direction being toward the Sun on the 21st of June, and the inclination being about $23\frac{1}{2}°$. Now, if we bear in mind that the Earth is spinning round once in twenty-four hours, we shall immediately see how it is we get day and night. The Sun can only light up that half of the Earth turned toward it; consequently, at any moment, one-half of our planet is in sunshine, the other in shade—the rotation of the Earth bringing each part in succession from sunshine to shade, and from shade to sunshine.

owe to each? 182. What does Fig. 41 represent? How is it that we get the suc

183. But it will be asked, "How is it that the days and nights are not always equal?" We answer, by reason of the inclination of the axis.

In the first place, the days and nights are equal all over the world on the 22d of March and the 22d of September, which dates are called the *vernal* and the *autumnal equinox* for that very reason—*equinox* being derived from two Latin words meaning *equal night.*

Now let us look at the small circle marked on the Earth—it is the arctic circle; and let us suppose ourselves living in Greenland, just within that circle. What will happen? At the *vernal equinox* (it will be most convenient to follow the order of the year) we find that circle half in light and half in shade. One-half of the twenty-four-hours (the time of one rotation), therefore, will be spent in sunshine, the other in shade: in other words, the day and night will be equal, as before stated. Gradually, however, as we approach the *summer solstice* (going from left to right), we find the circle coming more and more into the light, in consequence of the inclination of the axis, until, when we arrive at the solstice, in spite of the Earth's rotation we cannot get out of the light. At this time we see the midnight sun due north. The Sun, in fact, does not set.

The solstice passed, we approach the *autumnal equinox*, when again we shall find the day and night equal, as we did at the vernal equinox. But when we come to the *winter solstice*, we get no more midnight suns: as shown in the figure, all the circle is situated in the shaded portion; hence, in spite of the Earth's rotation, we cannot get out of the darkness, and we do not see the Sun even at noonday.

There will now be no difficulty in understanding how at the poles the years consist of one day of six months'

---

cession of day and night? 183. Explain the inequality of the days and nights, and the changes that occur in this respect as the Earth advances in her orbit.

duration, and one night of equal length. To comprehend our long summer days and short nights, we have only to take a point about half-way between the arctic circle and the equator, as marked on the plate, and reason in the same way as we did for Greenland. At the equator we shall find the day and night always equal.

184. Here is a table showing the length of the longest day in different latitudes, from the equator to the poles:—

°	′		Hours.	°	′		Hours.
0	0	(Equator)	12	65	48		22
16	44		13	66	21		23
30	48		14	66	32		24
41	24		15				Months.
49	2		16	67	23		1
54	31		17	69	51		2
58	27		18	73	40		3
61	19		19	78	11		4
63	23		20	84	5		5
64	50		21	90	0	(Pole)	6

185. What we have said about the northern hemisphere applies equally to the southern, but the diagram will not hold good, as the northern winter is the southern summer, and so on; moreover, if we could look upon our Earth's orbit from the other side, the direction of the motions would be reversed. The pupil should construct a diagram for the southern hemisphere for himself.

186. **The Change of Seasons.**—The changes to which we inhabitants of the temperate zones are accustomed, the heat of summer, the cold of winter, the medium temperatures of spring and autumn, depend simply upon the height which the Sun attains at mid-day—for the more nearly perpendicular the Sun's rays are, the more heat does the Earth absorb from them. This is proved by the

---

184. How long are the days and nights at the poles? At the equator? In what latitude is the length of the longest day 15 hours? Twenty hours? One month? Three months? 186. On what do the changes of season in the temperate

facts that on the equator the Sun is never far from the zenith—that is, the point directly overhead—and we have perpetual summer: near the poles, the Sun never gets very high, and we have perpetually the cold of winter. How, then, are the changing seasons in the temperate zones caused?

187. In Fig. 41 we were supposed to be looking down upon our system. We will now take a section from solstice to solstice through the Sun, in order that we may have a side view of it. Here, then, in Fig. 42, we have the Earth in two positions, and the Sun in the middle.

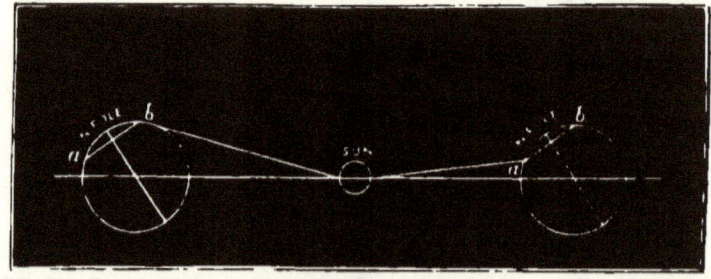

Fig. 42.—Explanation of the Apparent Altitude of the Sun in Summer and Winter.

On the left we have the Earth at the winter solstice, when the axis of rotation is inclined away from the Sun to the greatest possible extent. On the right we have it at the summer solstice, when the axis of rotation is inclined toward the Sun to the greatest possible extent. The line ab in both represents a parallel of latitude in the north temperate zone. The dotted line from the centre through b in the figure on the left, and through a in that on the right, shows the direction of the zenith—the direction in which our body points when we stand upright. We see that this line forms a larger angle with the line leading to the Sun—that is, the two lines open out wider—at the winter, than they do at the summer, solstice. Hence in

---

zones depend? How is this proved? 187. With Fig. 42, show that the Sun attains different heights at different seasons. How is the Earth's axis inclined

the latitude indicated the Sun is seen in winter at noon, low down, far from the zenith, while in summer it is nearly overhead.

The pupil should now make a similar diagram, to represent the position of the Sun at the equinoxes; he will find that the axis is not then inclined either to or from the Sun, but sideways,—the result being that the Sun itself is seen at the same distance from the point overhead in spring and autumn. Hence the temperature is nearly the same, though Nature apparently works very differently at these two seasons; in one we have seed-time, in the other the fall of the leaf.

FIG. 43.—THE EARTH, AS SEEN FROM THE SUN AT THE SUMMER SOLSTICE (Noon at London).

188. Perhaps the Sun's action on the Earth, in giving rise to the seasons, may be made clearer by inquiring how the Earth is presented to the Sun at the four seasons—that is, how the Earth would be seen by an observer at

at the equinoxes? What is the consequence, as regards the temperature? 188. What do Figs. 43 and 44 represent? What is the difference in the situation of

98 THE EARTH.

the Sun. First, then, for summer and winter. Figs. 43 and 44 represent the Earth as it would be seen from the Sun at noon in London, at the summer and winter solstices. In the former, England is seen well down toward the centre of the disk, where the Sun is vertical, or overhead; its rays are therefore most felt, and summer prevails. In the latter, England is near the northern edge of the disk,

FIG. 44.—THE EARTH, AS SEEN FROM THE SUN AT THE WINTER SOLSTICE
(Noon at London).

and farthest from the region where the Sun is overhead; the Sun's rays are consequently feeble, and winter reigns.

189. In Figs. 45 and 46, representing the Earth at the two equinoxes, we see that the position of England, with regard to the centre of the disk, is the same—the only difference being that in the two figures the Earth's axis is inclined in different directions. Hence there is no difference of temperature at these periods.

places in northern latitudes, in the two diagrams? 189. What do Figs. 45 and 46 represent? What is the only difference noticeable in these diagrams?

DIFFERENCE OF TIME. 99

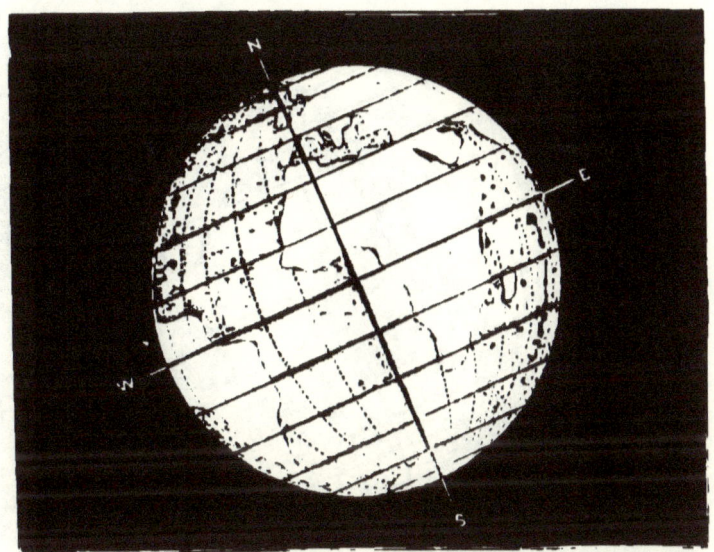

Fig. 45.—The Earth, as seen from the Sun at the Autumnal Equinox
(Noon at London)

190. Figs. 43, 44, 45, and 46, all represent London on the meridian which passes through the centre of the illuminated side of the Earth. It must therefore be noon at that place, as noon is half-way between sunrise and sunset. All the places represented on the western border have the Sun rising upon them; all the places on the eastern border have the Sun setting. As, therefore, at the same moment of absolute time we have the Sun rising at some places, overhead at others, and setting at others, we cannot have the same time, as measured by the Sun, at all places alike.

191. **Difference of Time and Longitude.**—In fact, as the Earth, whose circumference is divided into 360°, turns round once in twenty-four hours, the Sun appears to travel one twenty-fourth of 360°, or 15°, in one hour, from east to west. *One degree of longitude, therefore, makes a differ-*

---

190. How do these figures show that the time, as measured by the Sun, differs at different places? 191. What difference of time does one degree of longitude

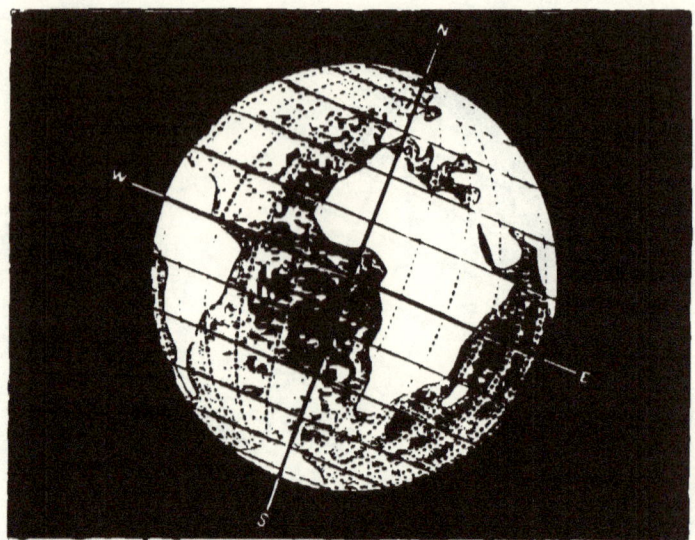

Fig. 46.—THE EARTH, AS SEEN FROM THE SUN AT THE VERNAL EQUINOX (Noon at London).

ence of four minutes of time, and vice versa,—the more easterly longitude having the later time.

The difference of longitude between New York and London being about 74°, the difference of time is 4 times 74 minutes, or 4h. 56m.—the time of London being later, because, being east of New York, the Sun comes sooner to its meridian. When, therefore, it is noon at New York, it is 56 minutes past 4 P. M. at London.

When it is noon at San Francisco, it is about 5 minutes after 3 P. M. at Philadelphia; required, their difference of longitude. Their difference of time being 3h. 5m., or 185 minutes, their difference of longitude will be as many times 1° as 4 minutes are contained times in 185 minutes, or 46¼°.

By this easy process navigators determine their longitude at sea. Taking with them a chronometer (an accurate watch) set according to the time of a given place (as, Greenwich or Washington), they ascertain the local time by observing with the sextant when the sun is at its high-

---

make? Why is this so? Of two places in different longitudes, which has the later time? When it is noon at New York, it is about 56 minutes past 4 P. M. at London; what is their difference of longitude? The difference of longitude between San Francisco and Philadelphia being about 46¼°, when it is noon at Philadelphia what time is it at San Francisco? How do navigators determine

est point; it is then noon. Reducing the difference of time to difference of longitude, as above, they find that they are so many degrees east or west of the meridian of the place for which their chronometer is set.

192. Structure of the Earth.—Having said so much of the motions of our Earth, let us now turn to its structure, or physical constitution.

We are all acquainted with the present appearance of our globe; we know that its surface is here land, there water; and that the land is, for the most part, covered with soil which permits of vegetation, varying according to the climate; while in some places meadows and wood-clad slopes give way to rugged mountains, which rear their bare or ice-clad peaks to heaven.

The first question that arises is, Was the Earth always as it is at present? The answer given by Geology and Physical Geography is, that the Earth was not always as we now see it, and that changes have been going on for millions of years, and are going on still.

193. The Earth's Crust.—It has been found that what is called the Earth's crust—that is, the outside of the Earth, as the peel is the outside of an orange—is composed of various *rocks* of different kinds and ages, all of them, however, belonging to two great classes:—

CLASS I. Rocks that have been deposited by water: these are called Stratified or Sedimentary Rocks.

CLASS II. Rocks that once were molten: these are called Igneous Rocks.

194. Stratified Rocks.—The stratified rocks have not always existed, for when we come to examine them closely it is found that they are piled one upon another in successive layers, as shown at the right of Fig. 48 below—the newer rocks lying on the older ones. The order in which

---

their longitude at sea? 192. Describe the present appearance of our globe. Was it always thus? 193. Of what is the Earth's crust composed? Into what classes are Rocks divided? 194. How are the Stratified Rocks arranged? Give a list

these rocks have been deposited, beginning with the uppermost, or those of latest formation, is as follows:—

Cainozoic, or Tertiary :
- Upper
  - Alluvium.
  - Drift.
  - Crag.
- Lower
  - Eocene.

Mesozoic, or Secondary :
- Upper
  - Cretaceous.
  - Oolite.
- Lower
  - Lias.
  - Trias.

Palæozoic, or Primary :
- Upper
  - Permian.
  - Carboniferous.
  - Devonian.
- Lower
  - Silurian.
  - Cambrian.
  - Laurentian.

195. That these beds have been deposited by water, and principally by the sea, is proved by two facts: First, that in their formation they resemble the beds being deposited by water at the present time; Secondly, that they nearly all contain the remains of fishes, reptiles, and shell-fish, in great abundance—indeed, some of the beds are composed almost entirely of the remains of animal life.

Fig. 47.—Fossil Fish.

Such remains, being dug out of the Earth, are called Fossils (from the Latin

---

of the stratified rocks in order, beginning with the latest. 195. How is it proved that these rocks have been deposited by water? What other name has been given to the stratified or sedimentary rocks? Why? What are Fossils? 196.

STRATIFIED ROCKS.

fossilis, dug). From their containing fossils, the stratified rocks have been called Fossiliferous.

196. It must not be supposed that the stratified rocks of which we have spoken are everywhere met with as they are shown in the table. Each bed could have been deposited only on those parts of the Earth's crust which were under water at the time; and since the earliest period of the Earth's history, volcanic action, earthquakes, and changes of level have been at work, as they are now—and much more effectively, either because the changes were more decided and sudden, or because they extended over immense periods of time.

It is found, indeed, that the stratified rocks have been upheaved and worn away again, bent and twisted to an enormous extent. Instead of being horizontal, as they must have been when they were originally formed at the bottom of the sea, they are now found in some cases tilted, as at the left of Fig. 48, and in others bent into irregular curves, as in Fig. 49.

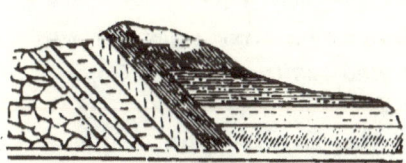
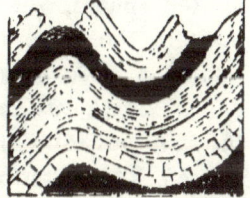

Fig. 48.—Stratified Rocks, tilted and horizontal. Fig. 49.—Stratified Rocks, curved.

Had this not been the case, the mineral riches of the Earth would forever have been out of our reach, and the surface of the Earth would have been a monotonous plain. As it is, although it has been estimated that the thickness of the series of stratified rocks, if found complete in any

---

What has interfered, in places, with the original arrangement of the stratified rocks? How are they sometimes found? What would be the thickness of the stratified rocks, if complete in one locality? What do we find with respect to each member of the series? What advantage results from this tilting? 197. How

one locality, would be 14 miles, each member of the series is found at the surface at some place or other.

197. **Igneous Rocks.**—The whole series of sedimentary rocks, from the most ancient to the most modern, have been disturbed by eruptions of volcanic materials, similar to those thrown up by Vesuvius and other volcanoes active in our own time, and intrusions of rocks of igneous origin from below. Of these igneous rocks, granite, which in consequence of its great hardness is largely used for paving and macadamizing, may here be taken as an example. These rocks are easily distinguishable from those of the first class, as they have no appearance of stratification and contain no fossils; their constituents are different and are irregularly distributed throughout the mass.

If we strip the Earth, then, in imagination, of the sedimentary rocks, we come to a kernel of rock, the constituents of which it is impossible to determine. It may, however, be supposed to be analogous to the older rocks of the granitic series, and to have been part of the original molten sphere, which must have been both hot and luminous, in the same way that molten iron is both hot and luminous. *Doubtless there was a time when the surface of our Earth was as hot and luminous as the surfaces of the Sun and stars are still.*

198. **The Interior of the Earth.**—Now, suppose we have a red-hot cannon-ball; what happens? The ball gradually parts with, or radiates away, its heat, and as it cools it ceases to give out light; but its centre remains hot long after the surface in contact with the air has cooled.

So precisely has it been with our Earth. We have numerous proofs that the interior of the Earth is at a high

---

have the whole series of stratified rocks been disturbed? What may be taken as an example of the Igneous Rocks? How are the igneous rocks distinguishable from the sedimentary? If we could strip the Earth of the sedimentary rocks, what should we come to? What was once doubtless the case respecting the Earth's surface? 198. What is the condition of the interior of the Earth? What

INTERIOR OF THE EARTH. 105

temperature at present, although its surface has cooled down. Our deepest mines are so hot that, without a perpetual current of cold fresh air, it would be impossible for the miners to live in them. The water brought up in artesian wells is found to increase in temperature 1 degree for from 50 to 55 feet of depth. Again, there are hot springs coming from great depths, the water of which is, in some cases, at the boiling-point—that is, 212° of Fahrenheit's thermometer. In the hot lava emitted from volcanoes we have further evidence of this interior heat, and that it is independent of the temperature at the surface; for among the most active volcanoes with which we are acquainted, are Hecla in Iceland, and Mount Erebus in the midst of the icy deserts which surround the south pole.

199. It has been calculated that the temperature of the Earth increases as we descend at the rate of 1° Fahrenheit in a little over 50 feet. We shall therefore have a temperature of

Fahr.		Miles.
212° or the temperature of boiling water	at a depth of about	2
750° or the temperature of red-hot iron	" "	$7\frac{1}{2}$
1,850° or the temperature of melted glass	" "	18
2,700° or the temperature at which everything we are acquainted with would be in a state of fusion	" "	28

200. If this be so, then the Earth's crust cannot exceed 28 miles in thickness—that is to say, the $\frac{1}{110}$th part of

---

evidences have we of the Interior heat? 199. What is the rate of increase of temperature, as we descend below the Earth's surface? At what depth would we have the temperature of boiling water? The temperature of red-hot iron? What temperature would we have at the depth of 18 miles? At the depth of 28 miles? 200. What follows, with respect to the thickness of the crust? What

the radius; so that it is comparable to the shell of an egg. But this question is one on which there is much difference of opinion, some philosophers holding that the liquid matter is not continuous to the centre, but becomes solid under the great pressure of the matter above. Indeed, evidence has recently been brought forward to show that the Earth may be a solid or nearly solid globe from surface to centre.

201. **Density of the Earth's Crust.**—The density of the Earth's crust is only about half of the mean density of the Earth taken as a whole. This has been accounted for by supposing that the materials of which it is composed are made denser at great depths than at the surface, by the enormous pressure of the overlying mass; but there are strong reasons for believing that the central portions are made up of much denser bodies than are common at the surface,—such as metals and the metallic compounds.

202. **The Flattening at the Poles explained.**—It was prior to the solidification of its crust, and while the surface was in a soft or fluid condition, that the Earth put on its present flattened shape, the flattening being due to a bulging out at the equator, caused by the Earth's rotation. If we arrange a thin flexible hoop, as shown in Fig. 50, so that the upper part of it may move freely up and down on an axis, and then make it revolve very rapidly, it will assume an oval form, bulging out at

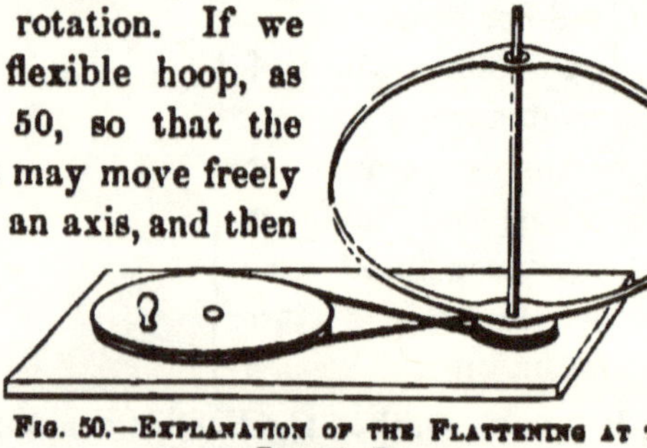

FIG. 50.—EXPLANATION OF THE FLATTENING AT THE EARTH'S POLES.

those parts which are farthest from the axis, the motion

---

opposite opinion is held by some? 201. How does the density of the Earth's crust compare with that of the whole planet? How is this accounted for? 202. How is the flattening at the Earth's poles accounted for? Illustrate this with

being there most rapid, just as the Earth does at the equator.

203. The form of the Earth, moreover, is exactly that which any fluid mass would take under the same circumstances. This has been proved by placing a quantity of oil in a transparent liquid of exactly the same density as the oil. As long as the oil was at rest, it took the form of a perfect sphere floating in the middle of the fluid, exactly as the Earth floats in space ; but the moment a slow rotary motion was given to the oil by means of a piece of wire forced through it, the spherical form was changed into a spheroidal one, like that of the Earth.

204. Thus the tales told by geology, the still heated state of the interior, and the shape of the Earth, agree; they all show that long ago the sphere was intensely heated and fluid.

205. **The Earth's Atmosphere.**—We now pass to the atmosphere, which may be likened to a great ocean, covering the Earth to a height not yet exactly determined. This height is generally supposed to be 45 or 50 miles, but there is evidence to show that we have an atmosphere of some kind at a height of 400 or 500 miles.

206. The atmosphere is the home of the winds and clouds, and it is with these especially that we have to do, in order to understand the appearances presented by the atmospheres of other planets. Although in any one place there seems to be no order in the production of winds and clouds, on the Earth considered as a whole there is the greatest regularity. The Sun's heat and the Earth's rotation on its axis are, in the main, the cause of all atmospheric disturbances.

Fig. 50. 203. What experiment, bearing on this flattening at the poles, has been made with oil ? 204. What is the conclusion drawn respecting the condition of the Earth long ago ? 205. To what may the atmosphere of the earth be likened ? How high does it extend ? 206. Of what is the atmosphere the home ? Is there any regularity in the production of winds and clouds ? What are the principal causes of atmospheric disturbances ? 207. As regards calms and winds, into

108 THE EARTH.

207. Belts of Calms and Winds.—If we examine a map showing the movements and conditions of the atmosphere, we shall find, encircling the Earth along the equator, a belt of *Equatorial Calms*. North of this we have the belt of *Trade-winds*, which blow from the north-east; on the south we have a similar belt where the prevailing winds are south-east.

Going from these belts toward the poles, we have on the north and south respectively the *Calms of Cancer* and the *Calms of Capricorn*. Still farther toward the Poles, we find the *Anti-trades*, blowing in the northern hemisphere from the south-west, and in the southern hemisphere from the north-west. At the poles there is a region of *Polar Calms*.

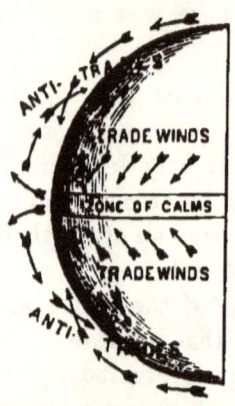

Fig. 51.—THE TRADES AND ANTI-TRADES.

208. Now, how are the winds just mentioned set in motion? The equatorial regions are the part of the Earth which is most heated; consequently the air there becomes rarefied and ascends, and a surface-wind sets in toward the equator on both sides to fill its place: these are the trade-winds. The air thus wafted toward the equator is soon itself heated and ascends, and accumulating in the higher regions flows as an upper current toward either pole. Thus are produced the anti-trades referred to above, which in the regions beyond the calms of Cancer and Capricorn descend to the Earth's surface. The equatorial belt (some 5° wide) in which the heated air is constantly ascending, is remarkable for daily rains, often accompanied with thunder and lightning.

What successive belts do we find the Earth's surface divided? 208. How are the trade-winds produced? How, the anti-trades? For what is the equatorial belt remarkable? 209. What makes the trade-winds deviate in direction from due

209. If the Earth did not turn on its axis, we should still have the trade-winds, but they would blow due north and south from the poles to the equator. Their direction is modified by the Earth's rotation. Coming from higher latitudes with the less rapid rotary motion which there belongs to the Earth's surface, to the equatorial regions which have a more rapid motion, the Earth, as it were, slips from under them toward the east; and the winds, lagging behind, though really themselves also moving eastward, *appear* to come *from* the east, forming north-east winds north of the equator, and south-east winds south of it.

In like manner, the anti-trades, endowed with the more rapid rotary motion of the equator, as they go toward the poles, arrive at regions where the rotary motion is less rapid. The Earth's surface, therefore, now lags behind, and the winds appear to blow, as they really do, toward the east, forming south-west winds in the northern hemisphere, and north-west winds in the southern.

210. It is the Sun, therefore, that sets all this atmospheric machinery in motion, by heating the equatorial regions of the Earth; and as the Sun changes its position with regard to the equator, crossing it twice in the course of the year, so do the calm-belt and trade-winds. The belt of equatorial calms follows the Sun northward from January to July, when it reaches $23\frac{1}{2}°$ N. lat., and then retreats, till at the next January it is in $23\frac{1}{2}°$ S. lat.

211. **Clouds, Rain, Snow, Hail.**—To the radiation from the Earth, combined with the fact that more or less watery vapor, or moisture, is always present in the air, must be ascribed the formation of mist and clouds, and the precipitation of rain, snow, and hail. The amount of

---

north and south? Explain how the Earth's rotation operates on them. How does the Earth's rotation affect the anti-trades? 210. How, and why, do the equatorial calm-belt and the trade-winds change their position? 211. To what are mist, cloud, rain, etc., due? Explain how clouds and rain are formed. Un-

moisture that the air can hold varies with its temperature; the warmer the air, the greater is its capacity for moisture. Hence, when air heavily charged with moisture derived by evaporation from the water-surfaces of the earth is chilled by a cold wind or contact with a mountain-side, its moisture is condensed into clouds or mist, and rain, snow, or hail, is formed. On the other hand, if a cloudy atmosphere is heated by the direct action of the Sun or by a current of warm air, its capacity for moisture is increased, and the clouds disappear.

212. **Chemical Elements of the Earth.**—We now come to the materials of which the Earth, including its atmosphere, is composed. These are 64 in number, and are called the Chemical Elements. They consist of

Non-metallic elements, or Metalloids. { Nitrogen, oxygen, hydrogen, chlorine, bromine, iodine, fluorine, silicon, boron, carbon, sulphur, selenium, tellurium, phosphorus.

Metallic elements. { *Metals of the alkalies:*—Potassium, sodium, cæsium, rubidium, lithium. *Metals of the alkaline earths:*—Calcium, strontium, barium. *Other metals:*—Aluminum, zinc, iron, tin, tungsten, lead, silver, gold, etc.

Of these 64 elements, combined with each other for the most part in various ways, the Earth and every object that we see around us are composed.

213. The elements which constitute the great mass of the Earth's crust are comparatively few—aluminum, calcium, carbon, chlorine, hydrogen, magnesium, oxygen, potassium, silicon, sodium, sulphur. Oxygen combines

---

der what circumstances will clouds disappear? 212. What is meant by the Chemical Elements? How many are there? Into what two classes are they divided? Name the non-metallic elements. Into what classes are the metallic elements subdivided? Name some of each class. 213. What elements constitute the

COMPOSITION OF THE AIR.

with many of these elements, and especially with the earthy and alkaline metals; indeed, about one-half of the Earth's crust is composed of oxygen in a state of combination. Thus sandstone, the most common sedimentary rock, is composed of silica, which is a compound of silicon and oxygen, and is half made up of the latter; granite, a common igneous rock, composed of quartz, felspar, and mica, is nearly half made up of oxygen in a state of combination in those substances.

214. **Composition of the Air.**—The chemical composition, by weight, of 100 parts of the atmosphere at present is as follows:—Nitrogen, 77 parts; Oxygen, 23 parts. Besides these two main constituents, we have

Carbonic acid, . quantity variable with the locality.
Aqueous vapor, . quantity variable with the temperature and humidity.
Ammonia, . . . a trace.

We said at present, because, when the Earth was molten, the atmosphere must have been very different. We had, let us imagine, close to the still glowing crust—consisting perhaps of acid silicates—a dense vapor, composed of compounds of the materials of the crust which were volatile only at a high temperature; the vapor of chloride of sodium, or common salt, would be present in large quantities; above this, a zone of carbonic acid gas; above this again a zone of aqueous vapor, in the form of steam; and lastly, the nitrogen and oxygen.

As the cooling went on, the lowest zone, composed of the vapor of salt and other chlorides, would be condensed on the crust, covering it with a layer of these substances in a solid state. Then it would be the turn of the steam to condense, and form water; this would fall on the layer

---

great mass of the Earth's crust? Of these, which enters most largely into the composition of matter? Of what is sandstone composed? Of what, granite?
214. What is at present the chemical composition of the air? Give an account of the atmosphere, when the Earth was molten. As the cooling went on, what

of salt, and dissolving it would form in time the ocean and seas, which would consequently be salt from the first moment of their appearance. Then, in addition to the nitrogen and oxygen which still remain, we should have the carbonic acid; this, in the course of long ages, was used up by its carbon going to form a luxurious vegetation, the remains of which are still to be seen in the coal that warms us and does nearly all our work.

215. It is the presence of vapor in our lower atmosphere that renders life possible. When the surface of the Earth was hot enough to prevent the formation of the seas, as the water would be turned into steam again the instant it touched the surface, there could be no life. Again, if ever the surface of the Earth be cold enough to freeze all the water and all the gaseous vapor in the atmosphere, life—as we have it—would be equally impossible.

216. **The Nebular Hypothesis**, before alluded to, here comes in and teaches that, prior to the Earth's being in a fluid state, it existed as part of a vast nebula, the parent of the Solar System; that this nebula gradually contracted and condensed, throwing off the planets one by one, some of which in turn threw off satellites; and that its central portion, condensed perhaps to the fluid state, exists at present as the glorious heat-giving Sun.

Although, therefore, we know that stars give out light because they are white-hot bodies, and that planets are not self-luminous because they are comparatively cold, we must not suppose that the planets were always cold, or that the stars will always be white-hot. There is good reason for supposing that all the planets were once white-hot, and gave out light as the Sun does now.

---

changes took place? What became of the carbonic acid? 215. To what is the presence of vapor in the atmosphere essential? Under what circumstances could there be no life? 216. What does the nebular hypothesis teach respecting the former condition of the Earth? What must we not suppose with regard to the planets and the stars? What is there good reason to suppose respecting the planets?

CHAPTER VI.

THE MOON.

217. **Size.**—The Moon, as already stated, is one of the satellites, or secondary bodies; and, although it appears to us at night to be infinitely larger than the fixed stars and planets, it is a little body but 2,153 miles in diameter. So small is it that 49 moons would be required to make one Earth, 1,245,000 earths being required, as we have seen, to make one Sun.

218. **Distance from the Earth.**—The apparent size of the Moon, then, must be due to its nearness. This we find to be the case. The Moon revolves round the Earth in an elliptical orbit, having the Earth at one of its foci, at an average distance of only 237,640 miles, which is equal to about 10 times round our planet. As the Moon's orbit is elliptical, she is sometimes nearer to us than at others. The greatest and least distances are 253,263 and 221,436 miles; the difference is 31,827 miles. When nearest us, of course she appears larger than at other times, and is said to be *in perigee* (περὶ, *near*, and γῆ, *the Earth*); when most distant, she is said to be *in apogee* (ἀπὸ, *from*, and γῆ, *the Earth*).

The Earth, by reason of its nearness, would of course look much larger to an observer on the Moon than any other of the heavenly bodies. When seen at the full, its disk would be as large as 13 full moons united would look to us. Bright spots would mark the continents, and the snow and ice about the poles; dark spots would indicate the water-surfaces; and variable ones, produced by the cloudy strata of the atmosphere, would at times be distinguishable.

---

217. To what class of heavenly bodies does the Moon belong? What is its size? What is its size, as compared with the Earth and Sun? 218. Why does the Moon look so large to us? What is the shape of its orbit? When is the Moon said to be *in perigee?* When, *in apogee?* What is its mean distance from the Earth? What appearance would the Earth present to observers on the

114   THE MOON.

219. **Period of Revolution.**—The Moon travels round the Earth in a period of 27d. 7h. 43m. 11½s. She requires more time to complete a revolution with respect to the Sun, which is called a Lunar Month, Lunation, or Synodic Period.

220. **Librations.**—The Moon, like the planets and the Sun, rotates on an axis; but there is this peculiarity in the case of the Moon, that her rotation and revolution round the Earth are performed in equal times. Hence we see only one side of our satellite. But, as the Moon's axis is inclined 83° 21' to the plane of its orbit, we sometimes see the region round one pole, and sometimes the region round the other. This is termed *Libration in latitude.*

There is also a *Libration in longitude*, arising from the fact that, though its rotation is uniform, its rate of motion round the Earth varies, so that we sometimes see more of the western edge, and sometimes more of the eastern. We have, moreover, a *daily Libration*, due to the Earth's rotation, carrying the observer to the right or left of a line joining the centres of the Earth and Moon. When on the right, or west, of this line, we should of course see more of the western edge of the Moon; when to the left, in the case of an eastern position, we should see more of the eastern edge.

By reason of these librations, instead of half the Moon's surface remaining constantly invisible, we see at one time or another about four-sevenths of its surface.

221. **Nodes.**—The plane in which the Moon performs her journey round the Earth is inclined 5° to the plane of

---

Moon, if there were any? 219. What is the period of the Moon's revolution? What is a Lunation? 220. How does the time of the Moon's rotation compare with that of her revolution? What follows? What is meant by the Moon's Libration in Latitude? What, by Libration in Longitude? What other libration is there? By reason of these librations, how much of the Moon's surface is at one time or another visible? 221. What angle do the plane of the Moon's orbit and the plane of the ecliptic form? What are the Nodes? By what names are the nodes distinguished? 222. What renders the motion of the Moon complicated? To what is its path round the Sun compared? What is said of the devi

THE MOON'S ORBIT. 115

Fig. 52.

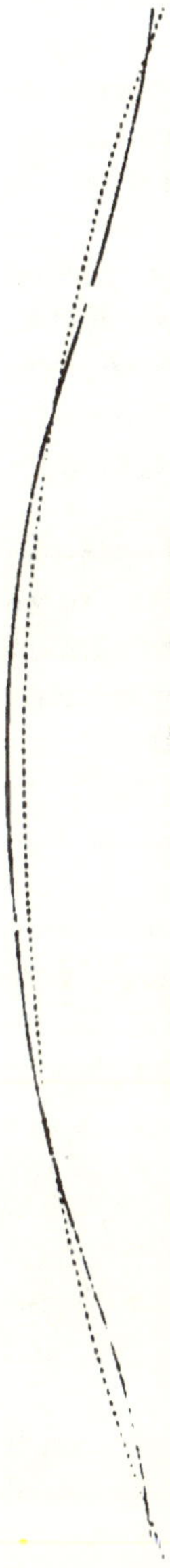

the ecliptic, in which the Earth performs her journey round the Sun (Art. 111). The two points in which the orbit of the Moon or any other celestial body intersects the Earth's orbit, are called the *Nodes*—that at which the body passes to the north of the ecliptic being distinguished as the Ascending Node, the other as the Descending Node. The line joining these two points is called the *Line of Nodes*.

222. **The Moon's Orbit.**—The Moon revolves round the Earth in an ellipse which has the Earth at one of its foci; but, while this revolution is going on, the Earth also is moving, in an elliptical orbit which has the Sun at one of its foci. Hence (leaving out of view the fact that the Sun also has a motion in which it is accompanied by both Earth and Moon) the motion of the Moon is quite complicated. We may get an idea of its path round the Sun, if we imagine a wheel going along a road to have a pencil fixed to one of its spokes, so as to leave a trace on a wall: such a trace would consist of a series of curves with their concave sides downward, and such is the Moon's path with regard to the Sun.

The total departure of the Moon from the Earth's orbit, however, does not exceed $\frac{1}{400}$ of the radius of the Earth's orbit; so that, unless drawn on a very large scale, the orbit of the Moon would appear to be identical with that of the Earth. In Fig. 52 the dotted line represents the Earth's orbit; the continuous line, the Moon's.

223. **Earth-shine.**—Besides the bright por-

---

tion of the Moon's path from that of the Earth? 223. What is meant by Earth-

tion lit up by the Sun, we sometimes see, in the phases which immediately precede and follow the New Moon, that the obscure part is faintly visible. This appearance is called the Earth-shine, and is due to that portion of the Moon reflecting to us the light it receives from the Earth. When this faint light is visible—when the "Old Moon" is seen "in the New Moon's arms"—the portion lit up by the Sun seems to belong to a larger moon than the other. This is an effect of what is called *irradiation*, and is explained by the fact that a bright object makes a stronger impression on the eye than a dim one, and appears larger the brighter it is.

224. **The Moon's Light.**—The average of four estimates gives the Moon's light as $\frac{1}{547,513}$ of that of the Sun; so we should want 547,513 full moons to give as much light as the Sun does. Now, there would not be room for so many in the half of the sky which is visible to us, as the full Moon covers $\frac{1}{210,000}$ of it; hence it follows that the light from a sky full of moons would not be so bright as sunshine.

225. **Apparent Difference of Size.**—At rising or setting, the Moon sometimes appears to be larger than it does when high up in the sky. This is a delusion, and the reverse of what we should expect; for, as the Earth is a sphere, we are really nearer the Moon by half the Earth's diameter when the part of the Earth on which we stand is beneath it than we are at moonrise or moonset, when we are situated, as it were, on the side of the Earth, halfway between the two points nearest to and most distant from the Moon. Let the student draw a diagram, and reason this out.

The larger appearance of the disk near the horizon is due simply to an error of judgment. It there seems placed

---

shine, and how is it produced? What difference of size is noticeable, and how is it explained? 224. How does the Moon's light compare in brightness with the Sun's? 225. When does the Moon appear largest? Why is this the reverse

beyond all the objects on the Earth's surface near the line along which we look, and therefore appears to be more distant than when it is overhead where there is no object near the line of view. Now, as it retains the same dimensions, and seems in the one case to be farther off than in the other, we intuitively endow it with a greater magnitude in the situation which is apparently the most remote —and it appears to the eye accordingly.

226. **Telescopic Appearance.**—A powerful telescope will magnify an object 1,000 times; that is to say, it will enable us to see it as if it were a thousand times nearer than it is. If the Moon were 1,000 times nearer, it would be about 240 miles off; consequently astronomers can see the Moon as if it were situated at this comparatively small distance, and they have studied and mapped with considerable accuracy the whole of the surface of our satellite which is turned toward us.

227. With the naked eye we see that some parts of the Moon are much brighter than others; there are dark patches, which, before large telescopes were in use, were thought to be oceans, gulfs, etc., and were so named. The telescope shows us that these dark markings are smooth plains, and that the bright ones are ranges of mountains and hill country broken up in the most tremendous manner by volcanoes of all sizes. A further study convinces us that the smooth plains are nothing but old sea-bottoms. In fact, once upon a time the Moon, like our Earth, was partly covered with water, and the land was broken up into hills and fertile valleys.

As on the Earth we have volcanoes, so had the Moon, with the difference that the size, number, and activity of the lunar volcanoes were far beyond any thing we can

---

of what we should expect? Explain why the Moon sometimes looks larger, when near the horizon. 226. How near does a powerful telescope bring the Moon? What have astronomers consequently been able to do? 227. What can we see with the naked eye, on the Moon's disk? What does the telescope show these dark patches and bright spots to be? What are the smooth plains found

imagine. Several of the craters exceed 50 miles in diameter, and one of them even measures 114½ miles,—beside which that of Kilauea, in the Sandwich Islands, the largest terrestrial crater known (2½ miles in diameter), dwarfs into insignificance.

228. The best way of seeing how the surface of our satellite is broken up in this manner, is to observe the *terminator*—as the boundary between the bright and shaded portions is called. Along this line the mountain-peaks are lighted up, while the depressions are in shade; and the shadows of the mountains are thrown the greatest distance on the illuminated portion. The heights of the mountains and depths of the craters have been measured by observing the shadows in this manner.

229. **The Crater-mountains and Ranges** of the Moon have been named after distinguished philosophers, astronomers, and travellers. Thirty-nine peaks have been found whose height exceeds that of Mont Blanc (15,870 feet). Dörfel is 26,691 feet high; the Ramparts of Newton, measured from the floor of the crater, are 23,853 feet high; Eratosthenes is 15,750 feet. These heights, it must be remembered, are much greater as compared with the size of the planet than the same elevations would be on the Earth's surface, as the Moon's diameter is but little more than one-fourth of the Earth's.

230. **The Crater Copernicus**, one of the most prominent objects in the Moon, is represented below. The details of the crater itself and of its immediate neighborhood reveal to us unmistakable evidences of volcanic action. The floor of the crater is strewn over with rugged masses, while outside the crater-wall (which on the left-hand side casts a shadow on the floor, as the drawing was taken

---

to be? What was once the case with respect to the Moon? Describe the lunar craters. 228. What is the best way of seeing how the surface of the Moon is broken up? 229. After whom have the crater-mountains and ranges of the Moon been named? How many peaks have been found higher than Mont Blanc? Mention some lunar peaks, and their height. 230. Describe the crater Coperni-

LUNAR CRATERS.

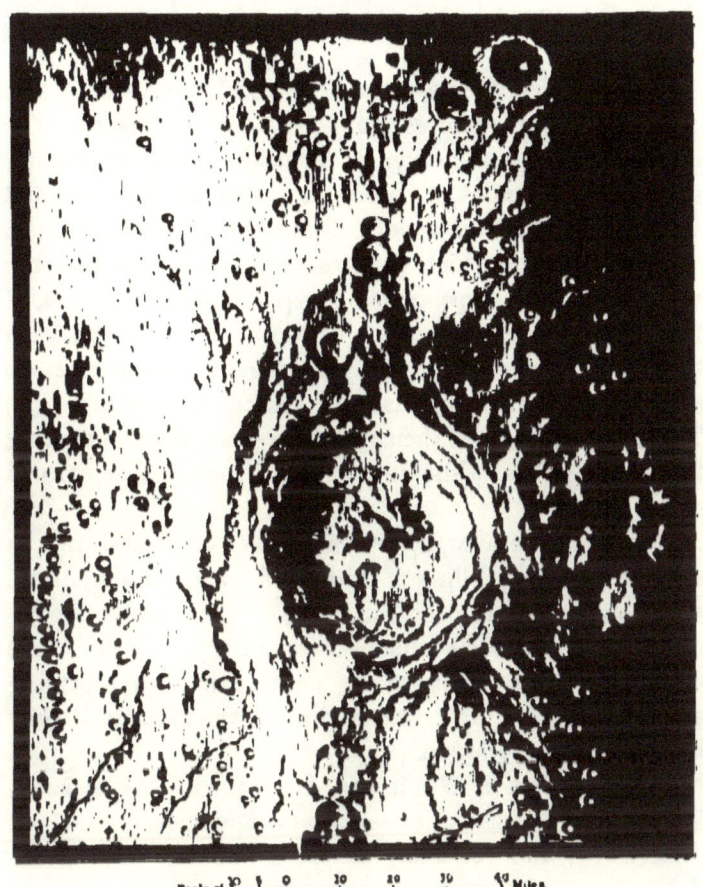

THE LUNAR CRATER COPERNICUS.

soon after sunrise at Copernicus, and the Sun is to the left) many smaller craters are distinctly visible, those near the edge forming a regular line. Enormous unclosed cracks and chasms are also distinguishable. The depth of the crater-floor, from the top of the wall, is 11,300 feet; and the height of the wall above the general surface of the Moon is 2,650 feet. The irregularities in the top of the wall are well shown in the shadow. The scale of miles attached to the drawing shows the enormous proportions of the crater.

231. **Walled Plains,** and curious markings called **Rilles,** are interesting features on the Moon's surface. The diameter of the walled plain Schickard, near the southeast limb or edge of the Moon, is 133 miles. Clavius and Grimaldi have diameters of 142 and 138 miles respectively.

The rilles, of which 425 are now known, are trenches with raised sides more or less steep. Besides the rilles, at full Moon, bright rays are seen, which seem to start from the more prominent mountains. Some of these rays are visible under all illuminations. Those emanating from Tycho are different in their character from those emanating from Copernicus, while those from Proclus form a third class.

Somewhat similar appearances have been produced on a glass globe by filling it with cold water, closing it up, and plunging it into warm water. This causes the enclosed cold water to expand very slowly, and the globe eventually bursts, its weakest point giving way, forming a centre of radiating cracks similar to the fissures, if they be fissures, in the Moon.

232. **Absence of Water and Atmosphere.**—As far as we know (with possibly one exception, which is not yet established) the volcanoes of the Moon are now all extinct; the oceans have disappeared, and no water exists on its surface; the valleys are no longer fertile; nay, the very atmosphere has apparently left our satellite, and that little celestial body which probably was once the scene of various forms of life now no longer supports them.

The absence of water and atmosphere may be accounted for by supposing that, on account of the small mass of the

---

cus. 231. What other interesting features are there on the Moon's surface? What is the diameter of the walled plain Schickard? Of Clavius? Of Grimaldi? What are the Rilles? How many rilles are known? What are seen apparently starting from the more prominent mountains? How have similar appearances been produced? 232. What is the present condition of the Moon's surface? How

Moon, its original heat has all been radiated into space. The cooling of its mass would be attended with contraction, and the formation of vast caverns in the interior, which would communicate by fissures with the surface. In these internal receptacles the ocean may have been swallowed up, to such a depth that not even the fierce heat of the long lunar day can draw it forth in the form of vapor. The Moon, then, may be a picture of what the Earth is destined to be—revolving round the Sun, an arid and lifeless wilderness—if ever its internal heat be wholly lost.

233. We say that the Moon has no atmosphere; (1) because we never see any clouds there, and (2) because, when the Moon gets between us and a star, the star disappears at once, and does not seem to linger on the edge, as it would do if there were an atmosphere.

The lunar days must have a singular aspect. There being no atmosphere to diffuse the solar light, the fiery disk of the Sun stands out sharp and distinct against the background of the sky, everywhere else dark except where it is dotted with stars. There is no cloud, no wind, no twilight, no sound—but everywhere a silent and lifeless desert.

234. The Moon rotates on her axis, as we do, only more slowly; hence the changes of day and night occur there as here. But instead of 24 hours, the Moon's day is $29\frac{1}{2}$ of our days long; so that each portion of the surface is in turn exposed to, and shielded from, the Sun for a fortnight. As there is no atmosphere either to shield the surface or to prevent radiation, it has been conjectured that the surface is, in turn, hotter than boiling water, and colder than any thing we have an idea of.

---

may the absence of water be accounted for? 233. Why do we say that the Moon has no atmosphere? What peculiarities must the lunar days present? 234. How long is the Moon's day? With what change of temperature must the alternation of day and night be attended? 235. What is meant by the Phases of the

122   THE MOON.

235. **Phases of the Moon.**—Let us now explain what are called the Phases of the Moon,—that is, the different shapes this luminary assumes. The Moon, like the Earth, gets all her light from the Sun. Now, it is clear that the Sun can only light up that half of the Moon which is turned toward it; it is equally clear that, if we were on the same side of the Moon as the Sun is, we should see the lit-up half—if we were on the other side, being opposite the dark half, we should see nothing at all of the Moon. Now, this is exactly what happens.

In Fig. 53 we suppose the plane of the Moon's orbit to correspond with the plane of the ecliptic, and the Sun to

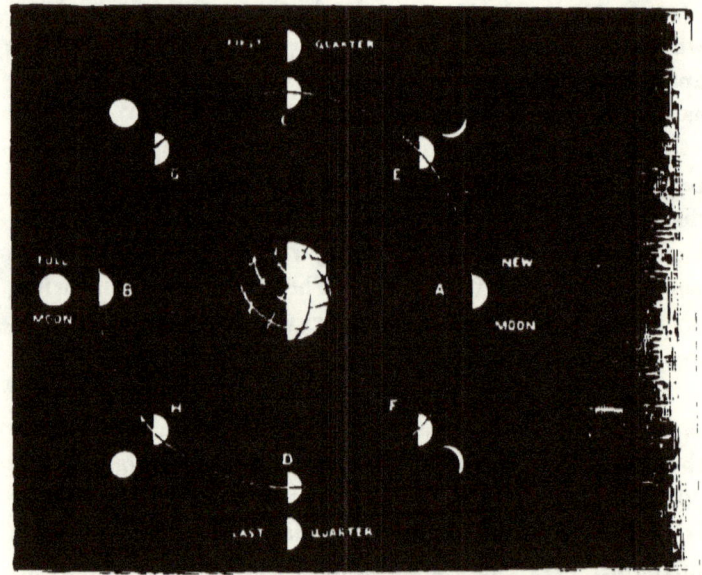

FIG. 53.—THE PHASES OF THE MOON.

lie to the right; the Moon's orbit is represented, and at its centre the Earth, the half turned toward the Sun being lighted up. When the Moon is at A, the illuminated side is away from the Earth, and we cannot see it; this is the

---

Moon? In explaining these, what facts must first be taken into consideration? With the aid of Fig. 53, explain the different phases. 236. Mention in order the

position occupied by the *New Moon*—and practically we do not see the New Moon at all. Now let us take the Moon at *B*; at this point we face the lit-up portion and see all of it. Now, this occurs at *Full Moon*, when the Moon arrives at such a position in her orbit that the Sun, the Earth, and herself, are in the same line, the Moon lying outside, and not between the other two as at the time of New Moon.

At *C* and *D* our satellite is represented midway between these two positions. At *C* one-half of the lit-up Moon is visible—that half lying to the right, as seen from the Earth; at *D* we see the illuminated portion lying to the left, looking from the Earth. These positions are those occupied by the Moon at the *First Quarter* and *Last Quarter* respectively. When the Moon is at *E* and *F*, we see but a small part of the illuminated portion, and have a *crescent Moon*, the convex side in both cases being turned to the Sun. At *G* and *H* the Moon is said to be *gibbous*.

236. We have, then, in succession, the following phases:—

New Moon. The Moon is invisible to us, because the Sun is lighting up one side and we are on the other.

Crescent Moon. We begin to see a little of the illuminated portion, but the Moon is still so nearly in a line with the Sun, that we only catch, as it were, a glimpse of the bright side, and that for a short time after sunset.

First Quarter. As seen from the Moon, the Earth and Sun are at right angles to each other. When the Sun sets in the west, the Moon is south; hence the right-hand side of the Moon is illuminated.

Gibbous Moon. The Moon is now more than half lighted up on the right-hand side.

Full Moon. The Earth is now between the Sun and Moon, and therefore the entire bright half of the Moon is visible.

From Full Moon we return, through similar phases in reversed order, to the New Moon, when the cycle recommences. So that, from New Moon the illuminated portion of our satellite *waxes*, or increases in size, till Full Moon, and then *wanes*, or diminishes, to the next New Moon; the illuminated portion, except at Full Moon, being separated from the dark one by a semi-ellipse, called, as we have seen (Art. 228), the Terminator.

CHAPTER VII.

ECLIPSES.

237. IN explaining the phases of the Moon, as represented in Fig. 53, we supposed the motion of this luminary to be performed in the plane of the ecliptic; but, as stated in Art. 221, this is not the case. If it were, every New Moon would put out the Sun; and as the Earth, like every body through which light cannot pass, casts a shadow, every Full Moon would be hidden in that shadow. Such phenomena are called Eclipses, and they do happen sometimes; let us see under what circumstances.

238. **Eclipses explained.**—One-half of the Moon's journey is performed above the plane of the ecliptic, one-half below it; hence at certain times—twice in each revolution

---

while the Moon is waning. How is the illuminated portion separated from the dark part?

237. If the Moon's orbit lay in the plane of the ecliptic, what would be the

EXPLANATION OF ECLIPSES. 125

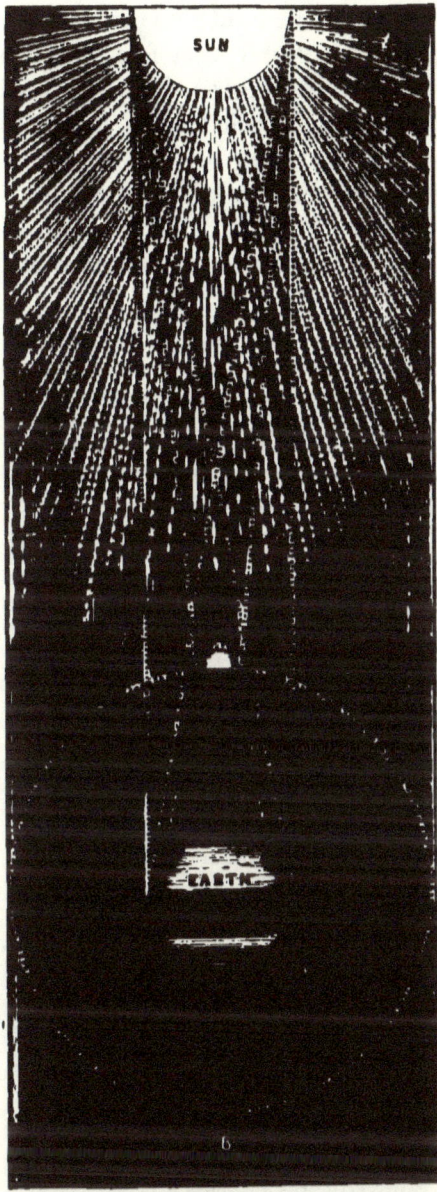

Fig. 54.—Explanation of Solar and Lunar Eclipses.

—the Moon is in that plane, at those parts of it called the *Nodes*. Now, if the Moon, when in either node, happens to be in line with the Earth and Sun, we have an eclipse. If the Moon is new, it is directly between the Earth and the Sun, and the Sun is eclipsed. If the Moon is full, the Earth is between it and the Sun, and the Moon is eclipsed. This will be made clear by the accompanying diagram.

In Fig. 54 we have the Sun and the Earth, and the Moon in two positions, *A* representing it as *new*, and *B* as *full*. The level of the page represents the plane of the ecliptic. We suppose, in both cases, that the Moon is at a node,—in the plane of the ecliptic, neither above nor below it.

At *A*, the Moon stops the Sun's light; its shadow falls on a part of the Earth, and the people, therefore, who live on that part cannot see the Sun, because the Moon is in the way. Hence we have what is called an *eclipse of the Sun*.

At *B*, the Moon is in the shadow of the Earth, and the Moon cannot receive any light from the Sun, because the Earth is in the way. Hence we have what is called an *eclipse of the Moon*.

239. It will be seen from the figure, that whereas the eclipse of the Moon by the shadow of the much larger Earth will be more or less visible to the whole side of the Earth turned away from the Sun, the shadow cast by the small Moon in a solar eclipse is, on the contrary, so limited that the eclipse is seen over a small area only.

240. **Umbra and Penumbra.**—In Fig. 54 two kinds of shadows are shown, one much darker than the other; the former is called the Umbra (a Latin word, meaning *shadow*), the latter the Penumbra (from the Latin *pœne*, almost, and *umbra*, a shadow). If the Sun were a point of light merely, the shadow would be all umbra; but it is so large, that round the umbra, within which no part of the Sun is visible, there is a belt where a portion of it can be seen; hence we get a partial shadow, which constitutes the penumbra. This will be made clear if we take two candles to represent any two opposite edges of the Sun, place them rather near together, at equal distances from a wall, and observe the shadow they cast on the wall from any object; on either side the dark shadow thrown by both candles, will be a lighter shadow thrown only by one.

241. **Total Eclipse of the Moon.**—In a total eclipse of

---

consequence? 238. What is meant by the Nodes? Under what circumstances do eclipses take place? Explain the occurrence of eclipses, with Fig. 54. 239. How do eclipses of the Moon and Sun compare, as regards the area over which they are respectively visible? 240. What is the difference between the umbra and the penumbra? How is the penumbra produced? How may the formation of these shadows in an eclipse be illustrated? 241. Describe the several steps

the Moon, as this body travels from west to east, we first see its eastern side slightly dim as it enters the penumbra; this is the *first contact with the penumbra*, spoken of in almanacs. At length, when the umbra is reached, the eastern edge becomes almost invisible, and we have the *first contact with the dark shadow*. The circular shape of the Earth's shadow is distinctly seen, and at last the Moon enters it entirely. Even then, however, the Moon's disk is scarcely ever wholly obscured; the Sun's light is bent by the Earth's atmosphere toward the Moon, and sometimes tinges it with a ruddy color.

A total eclipse of the Moon may last about $1\frac{3}{4}$ hours. When the Moon emerges from the umbra we have the *last contact with the dark shadow;* then follows the *last contact with the penumbra*, and the eclipse is over.

242. **Partial Eclipse of the Moon.**—If the Moon is very near, but not exactly at, a node, we have only a partial eclipse of the Moon, the degree of eclipse depending upon the distance from the node. For instance, if the Moon is north of the node, the lower limb may enter the upper edge of the penumbra or umbra; if south of the node, the upper portion may be obscured.

243. **Total Eclipse of the Sun.**—In a total eclipse of the Sun, the diameter of the shadow which falls on the Earth is never large, averaging about 150 miles. As the Moon, which throws the shadow, moves in its orbit from west to east, the eclipse always begins at the western edge of the Sun. The shadow first strikes the illuminated side of the Earth on the west, and sweeps eastward across it with great rapidity. The longest time an eclipse of the Sun can be total at any place is a little less than eight minutes; of course it is visible only at those places swept

---

In a total eclipse of the Moon. How long may such an eclipse last? 242. Under what circumstances have we a partial eclipse of the Moon? 243. What is the longest time an eclipse of the Sun can be total at any place? Explain why it is so short. What follows with respect to the number of total eclipses of the Sun?

by the shadow. Hence, in any one place total eclipses of the Sun are very rare; in London, for instance, prior to the total eclipse of 1715, no such phenomenon had been visible for a period of 575 years.

244. Annular Eclipse of the Sun.—When the Moon intervenes between the Sun and the Earth at such a distance from the latter as to make her apparent diameter less than the Sun's, a singular phenomenon is exhibited. The whole disk of the Sun is obscured except a narrow ring around the outside, encircling the darkened centre. This is called an Annular Eclipse (from the Latin *annulus*, a ring). Such an eclipse is represented in Fig. 55.

FIG. 55.—ANNULAR ECLIPSE OF THE SUN.

To explain an Annular Eclipse fully, we must give a few figures. As both Sun and Moon are round, or nearly so, the shadow from the latter is round; and as the Sun is larger than the Moon, the shadow ends in a point—its shape is, in fact, that of a cone, as shown in Fig. 55. Now, the length of this cone varies with the Moon's distance from the Sun, being of course shortest when the Moon is nearest to that luminary. The length of the Moon's shadow is about as follows:—

	Miles.
When the Moon and Sun are nearest together,	230,000
" " " farthest apart,	238,000

But the distance between the Earth and Moon varies as follows:—

	Miles.
When the Moon and Earth are nearest together,	221,436
" " " farthest apart,	253,263

Hence, when the Moon is farthest from the Earth, or in apogee, the shadow thrown by the Moon is not long enough to reach the Earth; at such times the Moon looks smaller than the Sun, and, if she be at a node, we have an Annular Eclipse.

245. Partial Eclipse of the Sun.—There may be Par-

---

244. What is an Annular Eclipse of the Sun? Under what circumstances is it produced? What is the shape of the Moon's shadow? Give figures to show that this shadow is not always long enough to reach the Earth. 245. Under

tial Eclipses of the Sun, for the same reason that we have partial eclipses of the Moon. As the Moon is not exactly at the node, in the one case she is not totally eclipsed, because she does not pass quite into the shadow of the Earth; and in the other, the Sun is not totally eclipsed, because the Moon does not pass exactly between us and the Sun.

246. To measure the extent of a partial eclipse, the diameter of the Sun or Moon, as the case may be, is divided into 12 equal parts, called *digits*, and the number of digits to which the greatest obscuration extends is stated.

247. **Recurrence of Eclipses.**—The nodes of the Moon are not stationary, but move backward upon the Moon's orbit, a complete revolution taking place with regard to the Moon in 18 years 219 days, nearly. The Moon in her orbit, therefore, meets the same node again before she arrives at the same position with regard to the Sun, one period being 27 d. 5 h. 6 m., called the *Nodical Revolution of the Moon;* and the other, 29 d. 12 h. 44 m., called the *Synodical Revolution of the Moon.* The node is in the same position with regard to the Sun after an interval of 346 d. 14 h. 52 m. This is called a *Synodic Revolution of the Node.*

Now, it so happens that nineteen synodical revolutions of the node, after which period the Sun and node would be alike situated, are equal to 223 synodical revolutions of the Moon, after which period the Sun and Moon would be alike situated. If, therefore, we have an eclipse at the beginning of the period, we shall have one at the end of it, the Sun, Moon, and node having returned to their original positions; and all the eclipses of the period (with an

---

what circumstances is a partial eclipse of the Sun produced? 246. How is the extent of a partial eclipse measured? 247. What motion have the nodes of the Moon? What is meant by the Nodical Revolution of the Moon? What is meant by its Synodical Revolution? What is the length of each of these periods? What is meant by the Synodic Revolution of a Node? How long a period is required for this revolution? Under what circumstances does a recurrence of

occasional exception) will recur in the same order and at about the same intervals as before. This period of 18 years 11 days 7 hours 40 min. 38 sec. (or, if 5 leap-years occur in the 18 years, 10 days instead of 11), was known to the ancient Chaldeans and Greeks under the name of *Saros*, and by its means eclipses were predicted before astronomy had made much progress.

248. **Phenomena attending a Total Eclipse of the Sun.**—A total eclipse of the Sun is at once one of the grandest and most awe-inspiring sights it is possible for man to witness. As the eclipse advances, but before the disk is wholly obscured, the sky grows of a dusky livid, or purple, or yellowish crimson, color, which gradually gets darker and darker, and the color appears to run over large portions of the sky, irrespective of the clouds. The sea turns lurid red. This singular coloring and darkening of the landscape is quite unlike the approach of night, and gives rise to strange feelings of sadness. The Moon's shadow sweeps across the surface of the Earth, and is even seen in the air; the rapidity of its motion and its intenseness produce a feeling that something material is rushing over the Earth at a speed perfectly frightful. All sense of distance is lost; the faces of men assume a livid hue, flowers close, fowls hasten to roost, cocks crow, birds flutter to the ground in fright, dogs whine, sheep collect together as if apprehending danger, horses and oxen lie down, obstinately resisting the whip and goad; in a word, the whole animal world seems frightened out of its usual propriety.

249. A few seconds before the commencement of the totality, the stars burst out, and surrounding the dark Moon on all sides is seen a glorious halo, generally of a silver-white light; this is called the **Corona**. It is slightly radiated in structure, and extends sometimes beyond the

---

eclipses take place? After how long an interval? What is this interval called? 248. Describe a total eclipse of the Sun, as regards its effect on the sky, the landscape, and the animal world. 249. What is the Corona? What are Aigrettes?

Moon to a distance equal to our satellite's diameter. Besides this, rays of light, called **Aigrettes,** diverge from the Moon's edge, and appear to be shining through the light of the corona. In some eclipses, parts of the corona have reached to a much greater distance from the Moon's edge than in others. It is supposed that the corona is the Sun's atmosphere, which is not seen when the sun itself is visible, owing to the overpowering light of the latter.

250. Sometimes when the advancing Moon has reduced the Sun's disk to a thin crescent, or in the case of an annular eclipse to a narrow ring, a peculiar notched appearance is presented in a part of the narrow strip, which makes it look like a string of beads (see Fig. 56). This phenomenon has been called "Baily's Beads," from the astronomer Baily, who was the first to describe it. It is supposed to be the effect of irradiation.

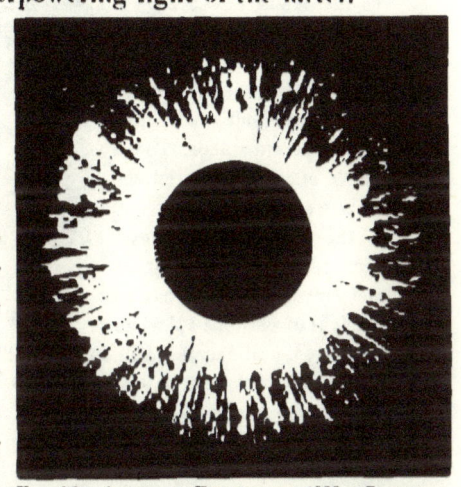

Fig. 56.—ANNULAR ECLIPSE OF 1836.—CORONA.— "BAILY'S BEADS."

251. When the totality has commenced, close to the edge of the Moon, and therefore within the corona, are observed fantastically-shaped masses, bright red fading into rose-pink, variously called **Red-flames** and **Red-prominences.** Fig. 57 shows this phenomenon, as exhibited in the total eclipse of 1860. Two of the most remarkable of these prominences hitherto noticed, were observed in the

---

What is the corona supposed to be? 250. What is meant by "Baily's Beads"? What is supposed to produce them? 251. When the totality has commenced, what are sometimes observed within the corona? Describe these red-flames, as

eclipse of 1851; they were described by Mr. Dawes as follows:—

"A bluntly triangular pink body was seen *suspended*, as it were, in the corona. This was separated from the Moon's edge when first seen, and the separation increased as the Moon advanced. It had the appearance of a large conical protuberance, whose base was hidden by some intervening soft and ill-defined substance. To the north of this appeared the most wonderful phenomenon of the whole: a red protuberance, of vivid brightness and very deep tint, arose to a height of perhaps 1¼' when first seen, and increased in length to 2' or more, as the Moon's progress revealed it more completely. In shape it somewhat resembled a *Turkish cimeter*, the northern edge being convex, and the southern concave. Toward the apex it bent suddenly to the south, or upward, as seen in the telescope. To my great astonishment, this marvellous object continued visible for about five seconds, as nearly as I could judge, after the Sun began to reappear."

FIG. 57.—LUMINOUS PROMINENCES observed at Rivabellosa, Spain, during the total eclipse of the Sun, July 18th, 1860.

252. It is certain that these prominences belong to the Sun, as those at first visible on the eastern side are gradually obscured by the Moon, while those on the western are becoming more visible, owing to the Moon's motion from west to east over the Sun. The height of some of them above the Sun's surface is upward of 40,000 miles. It is thought that they are incandescent clouds floating in the Sun's atmosphere, or resting upon the photosphere; but this has not as yet been definitely established.

253. **Number of Eclipses.**—In the Saros, or eclipse-

---

observed by Mr. Dawes in the eclipse of 1851. 252. What reason is there for supposing that these prominences belong to the Sun? What is the height of some of them? What are they thought to be? 253. How many eclipses occur

period of 18 years 11 days, there usually happen 70 eclipses, of which 41 are solar and 29 lunar. In any one year the greatest number that can occur is 7, and the least 2; in the former case, 5 of them may be solar, and 2 lunar; in the latter, both must be solar. Under no circumstances can there be more than 3 lunar eclipses in one year, and in some years there are none at all. Though eclipses of the Sun are more numerous than those of the Moon, in the proportion of 41 to 29, yet at any given place more lunar eclipses are visible than solar; because, while the former are visible over an entire hemisphere, the latter are seen only in a narrow strip which cannot exceed 180, and is usually about 150, miles in breadth.

254. **Memorable Eclipses.**—Thales, of Miletus, one of the seven wise men of Greece, was the first to give the true explanation of eclipses. He predicted a total eclipse of the Sun, which took place 585 b. c., and is memorable for having put an end to an engagement between the Medes and Lydians. Herodotus tells us that the day was suddenly turned to night, and that, when the contending armies observed the strange phenomenon, they ceased fighting and concluded a peace which was cemented by a twofold marriage.

Another total eclipse of the Sun, which occurred March 1st, 557 b. c., led to the capture of the Median city Larissa by the Persians, its defenders having withdrawn from its walls in alarm.

255. **Effects of Eclipses on the Uneducated.**—Though the cause of eclipses was understood by the wise men of antiquity, the people generally, as indeed the uneducated in modern times have done until quite recently, regarded

---

In the Saros? What is the greatest, and what the least, number that can occur in any one year? How many lunar eclipses may occur in a year? Which is oftener eclipsed, the Sun or the Moon? Why are there more lunar than solar eclipses visible at any given place? 254. Who was the first to give the true explanation of eclipses? What eclipse did Thales predict? What made this eclipse memorable? What other memorable eclipse is mentioned? 255. What

these phenomena with dread. Savage nations, not unnaturally, look upon them as omens of evil, and connect various superstitions with their occurrence.

The Hindoos, when they see the black disk of our satellite advancing over the Sun, believe that the jaws of a dragon are gradually eating it up. To frighten off the devouring monster, they commence beating gongs and rending the air with discordant screams of terror and shouts of vengeance. For a time their efforts have no effect; the eclipse still progresses. At length, however, the uproar terrifies the voracious dragon; he appears to pause, and, like a fish that has nearly swallowed a bait and then rejects it, he gradually disgorges the fiery mouthful. When the Sun is quite clear of the monster's jaws, a shout of joy is raised, and the exultant natives congratulate themselves on having, as they suppose, saved their deity from a disastrous fate. Elsewhere in India, the natives immerse themselves in the rivers up to the neck, which they regard as a most devout position, and thus seek to induce the luminary which is in process of eclipse to defend itself against the devouring dragon.

An eclipse of the Moon, March 1st, 1504, proved of great service to Columbus. During one of his voyages of exploration, he was wrecked on the coast of Jamaica. Reduced to the verge of starvation by the want of provisions, which the natives refused to supply, he took advantage of their ignorance of astronomy to save himself and his men. Knowing that an eclipse of the Moon was about to take place, he called the natives around him on the morning before its occurrence, and informed them that the Great Spirit was displeased because they had not treated the Spaniards better, and would shroud his face from them

---

has been the effect of eclipses on those unacquainted with their cause? What superstition do the Hindoos connect with a solar eclipse? What position do the natives assume in some parts of India? How did Columbus once save himself and his men from great straits?

that night. When the Moon became dark, the Indians, convinced of the truth of his words, hastened to him with plentiful supplies, praying that he would beseech the Great Spirit to receive them again into favor.

CHAPTER VIII.

THE INFERIOR AND SUPERIOR PLANETS.

256. To distinguish the planets which travel round the Sun within the Earth's orbit, from those which lie beyond it, the former, *i. e.*, Mercury and Venus, are termed **Inferior Planets**: while the latter, *i. e.*, Mars, Jupiter, Saturn, Uranus, and Neptune, are called **Superior Planets**. We proceed to consider these planets in turn.

257. **Mercury** (☿).—The nearest planet to the Sun whose existence is positively known, is Mercury. Under favorable circumstances, Mercury may be seen at certain times of the year for a few minutes after sunset, and then after an interval of some days for a few minutes before sunrise. At other times it keeps so close to the Sun as to be invisible, being lost in the superior brightness of his rays in the daytime, and setting and rising so nearly at the same time with him as to afford no opportunity of observation. It is never more than 29° distant from the Sun.

To the naked eye Mercury looks like a star of the third magnitude, twinkling (unlike the other planets) with a pale rosy light. Viewed through the telescope, it exhibits similar phases to those of the Moon (from full to new);

---

256. Into what two classes are the planets (the Earth excepted) divided? Name the Inferior Planets. Name the Superior Planets. 257. Which is the nearest Planet to the Sun? At what times is Mercury visible? Why is it not visible at other times? What is its greatest distance in degrees from the Sun?

this is because we see more of its illuminated side at one time than another. Fig. 58 shows the phases of Mercury when seen after sunset. Its phases when seen before sunrise are the same in reverse order, the illuminated part being turned in the opposite direction.

FIG. 58.—PHASES OF MERCURY, AND ITS COMPARATIVE SIZE AS SEEN AT DIFFERENT TIMES.

258. Mercury's orbit is more elongated than that of any other of the principal planets; it differs so much from a circle that at perihelion the planet is more than 14 millions of miles nearer the Sun than at aphelion. It follows from this, and from the fact that Mercury is sometimes between us and the Sun and sometimes on the other side of the Sun, that the distance of Mercury from the Earth differs greatly at different times. The apparent diameter of Mercury as seen from the Earth varies accordingly, being at the planet's nearest point $2\frac{1}{2}$ times as great as when it is farthest removed. This difference of size is represented in Fig. 58.

259. The mean solar heat received at Mercury is nearly 7 times as great as that of the Earth. The mean intensity of its light is also 7 times as great as ours, and the Sun seen from its surface would look 7 times as large as it does to us. We say its *mean* heat and light, for when it

---

How does Mercury look to the naked eye? What phases does it present, through the telescope? 258. Describe Mercury's orbit. How is it that Mercury's distance from the Earth varies so much? What change is noticeable in its apparent diameter? 259. How do the heat and light received at Mercury compare with

is nearest to the Sun the intensity of its heat and light is 10 times as great as ours, and when most distant only 4½ times as great. Hence the differences of temperature at different seasons are extreme; the seasons, also, are of very unequal duration. Every six weeks on an average there is a change of temperature nearly equal to the difference between frozen quicksilver and melted lead.

260. Mercury turns on its axis in 24 hours 5½ minutes, and its year comprises about 88 of our days. The inclination of its equator to the plane of its orbit is believed to be considerably greater than the Earth's; if this be so, the relative length of the days and nights must vary more than on our planet.

Mercury is the densest of the planets, having a specific gravity ¼ greater than that of the Earth. The force of gravity on its surface is about ⅜ that on the Earth's surface. The flattening at the poles is much greater than that of our planet, the polar diameter being to the equatorial as 28 to 29.

261. The nearness of Mercury to the Sun prevents us from obtaining any very accurate knowledge of its surface. There are indications, however, of the existence of mountains, one of which, in the southern hemisphere, is estimated to be over 11 miles high. There are also evidences of an atmosphere.

262. **Venus** (♀).—The second planet from the Sun is Venus. On account of its nearness, it appears larger and more beautiful to us than any other member of our planetary system. So bright is Venus that it is sometimes visible by day to the naked eye, and at night in the absence of the Moon casts a perceptible shadow. Viewed through

---

those of the Earth? 260. How long are Mercury's day and year? What is believed to be the case respecting the inclination of its axis? What would follow? How does Mercury compare with the Earth in density, and the force of gravity on its surface? How does its polar diameter compare with its equatorial? 261. Of what are there indications on the surface of Mercury? 262. Which is the second planet from the Sun? Describe the appearance of Venus. What

the telescope, it presents phases similar to those of Mercury (see Fig. 58) and the Moon. When seen as a crescent, owing to its nearness to the Earth at that time, its apparent diameter is nearly 6¼ times as great as when it is farthest off.

263. Venus is always below the horizon at midnight. During part of the year, it rises before the Sun, and ushers in, as it were, the day; when appearing at this time, the ancients styled it Phosphor or Lucifer (the light-bearer), and we call it the Morning Star. A few days after it ceases to be visible in the morning, it appears after sunset; it was then styled Hesperus or Vesper by the ancients, and is distinguished by us as the Evening Star. The greatest distance it attains from the Sun is 47°.

264. In size, density, and the force of gravity on its surface, Venus differs but little from the Earth. No flattening at the poles has been observed, from which it is inferred that in this respect also it resembles the Earth, as the flattening on our planet is so small that it would be imperceptible to an observer on Venus. In consequence of the nearly circular form of its orbit, its four seasons are nearly uniform in length; but the great inclination of its equator to the plane of its orbit (49° 58′) must, as in the case of Mercury, make a great difference in the relative length of day and night, and subject its polar regions to extreme changes of temperature. Venus's day is about 23¼ hours long, and its year is equal to about 224⅔ of our days.

265. Venus's heat and light are twice as intense as ours. A dense, cloudy atmosphere is believed to envelop the planet. Spots have been observed on its surface; and irregularities are seen in the terminator, which are sup-

posed to indicate lofty mountains, in some cases exceeding 20 miles in height.

266. **Mars** (♂).—Mars, fourth in order from the Sun, is the nearest to the Earth of the superior planets. Its day is of nearly the same length as ours; its year is about twice as long as our year; its diameter is two-thirds that of the Earth. The inclination of its axis to the plane of its orbit does not differ much from the Earth's, and its seasons are therefore similar to ours. When it is nearest to the Earth, its apparent diameter is 7 times as large as when it is farthest off. Mars has two moons, discovered in 1877.

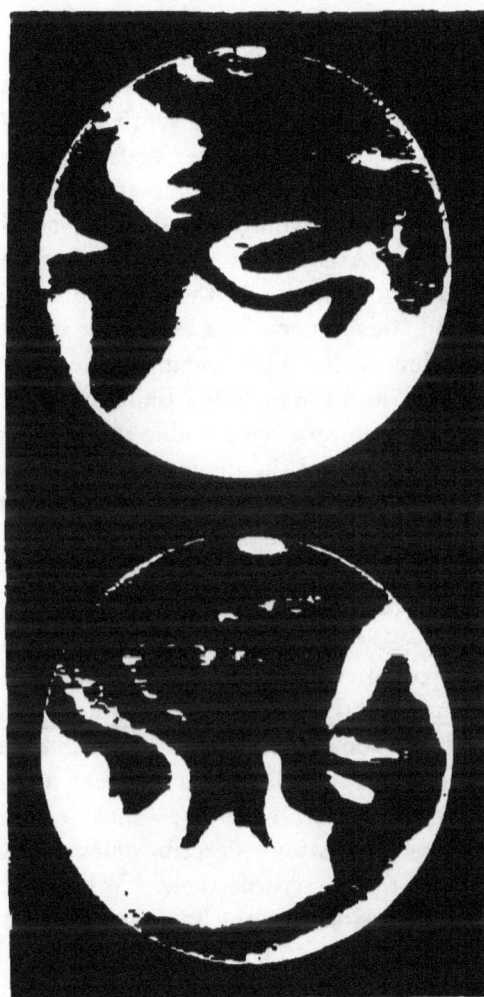

FIG. 59.—MARS IN 1862.

ties in the terminator indicate? 266. Which of the superior planets is nearest to the Earth? How does Mars compare with the Earth, as regards its day, its year, its diameter, and its seasons? How and why does its apparent diameter

140 THE INFERIOR AND SUPERIOR PLANETS.

267. In Fig. 59 are presented two sketches of Mars. Here we have something strangely like the Earth. The shaded portions represent water, the lighter ones land, and the bright spot at the top of the drawings is probably snow lying round the south pole.

The two drawings represent the planet as seen in a telescope, which inverts objects, so that the south pole of the planet is shown at the top. In the upper drawing, which was made on the 25th of September, a sea is seen on the left, stretching down northward; while, joined to it, as the Mediterranean is to the Atlantic, is a long, narrow sea, which widens at its termination. In the lower drawing, made September 23d, this narrow sea is represented on the left. The coast-line on the right reminds one of the Scandinavian peninsula, and the included Baltic Sea.

268. It will now be seen how we are able to determine the length of a planet's day and the inclination of its axis. We have only to watch how long it takes one of the spots near the equator to pass from one side to the other, and the direction in which it moves, to get at both these facts.

269. For a reason that will be understood when we come to deal with the effect of the Earth's revolution round the Sun on the apparent positions and aspects of the planets, we sometimes see the north pole of Mars, sometimes its south pole, and sometimes both. When but one pole is visible, the features which appear to pass across the planet's disk in about twelve hours—that is, half the period of the planet's rotation—describe curves with the concave side toward the visible pole. When both poles are visible they describe straight lines, exactly as in the case of the Sun (Art. 110). These changes enable all the surface to be seen at different times, and maps of Mars

---

vary? 267. Describe the surface of the planet, as represented in two sketches taken in 1862. 268. How can we determine the length of a planet's day and the inclination of its axis? 269. What is the direction of the features which cross the disk of Mars? How is the exact position of these features, as laid down in

have been constructed, the exact position of the features of the planet being determined by their latitude and longitude, as in the case of the Earth.

270. Mars has not only land, water, and snow, like the Earth, but also clouds and mists, and these have been watched at different times. The land is generally reddish when the planet's atmosphere is clear; this is due to the absorption of the atmosphere, as is the color of the setting Sun with us. Hence the fiery red light by which Mars is distinguished in the heavens. The water appears of a greenish tinge.

271. Now, if we are right in supposing that the bright spot surrounding the pole is ice and snow, we ought to see it rapidly decrease in the planet's summer. This is actually found to be the case, and the rate at which the thaw takes place is one of the most interesting facts to be gathered from a close study of the planet. In 1862, this decrease was very visible. The summer solstice of Mars occurred on the 30th of August, and the snow-zone was observed to be smallest on the 11th of October, or forty-two of our days after the highest position of the Sun. This very rapid melting may be ascribed to the inclination of the planet's axis, the great eccentricity of its orbit, and the fact that the summer of the southern hemisphere occurs when the planet is near perihelion.

272. Though we see in Mars so many things that remind us of the Earth, and show us that the extreme temperatures of the two planets are not far from equal, in one respect they differ widely. In consequence of the great eccentricity of the orbit of Mars, the seasons are not so nearly equal in length as with us, and owing to the longer

---

maps, determined? 270. How is Mars distinguished in the heavens? To what is this red light attributable? What tinge has the water on the surface of Mars? 271. What is the bright spot observed near the south pole of Mars supposed to be? What interesting fact was observed in connection with this spot in 1862? To what may the rapid thaw be ascribed? 272. In what respect does Mars differ widely from the Earth? What is the length of the several seasons in the

year, they are of much greater extent. In the northern hemisphere of the planet,

	days.	hrs.		days.	hrs.
Spring lasts	191	8	Autumn lasts	140	8
Summer "	181	0	Winter "	147	0

As we must reverse the seasons for the southern hemisphere, spring and summer, taken together, are 76 days longer in the northern hemisphere than in the southern.

273. **Jupiter** (♃).—Passing over the asteroids, which will be considered by themselves in the next chapter, we come to Jupiter, by far the largest planet of our system. Jupiter exceeds the Earth in bulk nearly 1,400 times. Its revolution round the Sun is performed in about 12 of our years, with a velocity 80 times as great as that of a

FIG. 60.—COMPARATIVE SIZE OF JUPITER AND THE EARTH.—BELTS OF JUPITER.

cannon-ball. It turns on its axis in less than 10 hours. The flattening at its poles is still greater than that at Mercury's, its polar diameter being to its equatorial diameter as 16 to 17. This flattening, represented in Fig. 60, is what we should expect from its very rapid rotation (Art. 202).

274. Jupiter stands nearly upright in its orbit, the inclination of its axis being only about 3°. Hence in any given part of the surface there is very little change of season. In the equatorial regions, summer reigns throughout the year; the temperate zones rejoice in perpetual spring; while around the poles winter continually prevails. The polar day is 6 years long, and is followed by a night of equal length. Owing to the deviation of its orbit from a circle, the planet is 46 millions of miles nearer the Sun at perihelion than at aphelion, and receives in consequence $\frac{1}{4}$ more heat.

275. Jupiter appears to the naked eye like a star of the first magnitude, bright enough sometimes, when the Moon is absent, in spite of its great distance, to cast a shadow like Venus. A glance at its disk, as represented in Fig. 60, shows us that we have something very unlike Mars. In fact it is surrounded by an atmosphere so densely laden with clouds, that of the actual surface we know nothing.

276. What are generally known as the *belts* of Jupiter are dusky streaks which cross a brighter background in directions generally parallel to the planet's equator. They are sometimes seen in large numbers, and extend almost to the poles. The largest belts are, for the most part, situated on either side of the equator, just as the two belts of trade-winds on the Earth lie on either side of the belt of equatorial calms. Outside these, again, we have the calms of Cancer and Capricorn represented, although these

---

its shape to have been affected by its rapid rotation? 274. What is the inclination of Jupiter's axis? What follows, respecting the seasons? How much more heat does the planet receive at perihelion than at aphelion? 275. How does Jupiter appear to the naked eye? By what is Jupiter surrounded? 276. Describe the belts of Jupiter, and their relative positions. What tint has the equatorial

belts are not so regularly seen, the portion of the planet's surface in which they are sometimes visible being liable to great changes of appearance, in a comparatively short time. The portions of the atmosphere representing the terrestrial calm-belts sometimes exhibit a beautiful rosy tint, the equatorial one especially.

Besides the belts, spots are seen, sometimes bright and sometimes dark, which have enabled us to determine the period of the planet's rotation. Its rotary velocity is so great that on the equator an observer would be carried round at the rate of 467 miles a minute, instead of 17 as on the Earth. This rapid rotation would necessarily break the cloudy surface into belts more than in our case or that of Mars. In the latter planet, indeed, no trace of cloud-belts has as yet been detected; their absence is perhaps due to its slow rotation and small size.

277. Though all astronomers do not agree that the surface of the planet is never seen, there are many good reasons for supposing this to be the case. In the first place, Mars and the Earth, whose atmospheres are nearly alike, have nearly the same density (Art. 157); while the density of Jupiter (and the same reasoning applies to Saturn, whose belts, as far as we can observe them, resemble Jupiter's), *if what we see is all planet*, is only about one-fifth that of the Earth, or not far from that of water. Now, there is no reason to suppose that the matter of which Jupiter is composed differs so very widely in character from that of Mars and the Earth. It seems more probable that the apparent volume of Jupiter (and, in like manner, that of Saturn) is made up of a large shell of cloudy atmosphere and a kernel of planet, and that the density of the real Jupiter (and the real Saturn) may not differ very much from that of the Earth.

---

belt? What are seen besides the belts? What is the planet's rotary velocity? What is the necessary effect of this rapid rotation on the atmosphere? To what is the absence of cloud-belts in Mars due? 277. What reason is there for sup-

Moreover, our own planet was most probably enveloped in a great shell of cloudy atmosphere, in one of the early stages of its history, before its crust had cooled down (Art. 214). The future ocean of Jupiter may now in like manner be spread about him, in the form of a blanket of cloud, 20,000 miles or more in thickness.

278. Besides the changing features of Jupiter itself, the telescope reveals to us four moons, which, as they course along rapidly in orbits lying nearly in the plane of the planet's orbit, lend additional interest to the picture. In

Fig. 61.—JUPITER AND HIS FOUR MOONS.

their various positions in their orbits, the satellites sometimes appear at a great distance from their primary; sometimes they come between us and the planet, appearing now as bright and now as dark spots on its surface. At other times they pass between the planet and the Sun, throwing their shadows on the planet's disk, and causing, in fact, eclipses of the Sun. They also enter the shadow cast by the planet, and are therefore eclipsed themselves; and sometimes they pass behind the planet, and are said to be *occulted*. Of these appearances we shall have more to say hereafter.

The passage of either a satellite or shadow is called a Transit. In a solar eclipse, could we observe it from Ve-

---

posing that Jupiter and Saturn are enveloped in immense shells of cloudy atmosphere? In what was our own planet probably once enveloped? 278. How many moons has Jupiter? In what different positions does the telescope exhibit them? What is meant by a Transit? 279. How do Jupiter's moons compare

nus, we should see the shadow of our Moon sweeping over the Earth's surface.

279. Referring to the sizes of these satellites and their distances from the planet, in Table III. of the Appendix, we find that all but one are larger than our Moon, and that all are farther from their primary than our Moon is from us. Like our Moon, they rotate on their axes in the same time as they revolve round their primary. This is inferred from the fact that their light varies, and that they are always brightest and dullest in the same positions with regard to Jupiter and the Sun.

280. **Saturn** (♄).—Saturn, which is next to Jupiter in distance from the Sun, is also next to it in size, having a volume about 750 times that of the Earth. Its day is not half so long as ours, but it is $29\frac{1}{2}$ of our years in making one complete revolution in its orbit.

FIG. 62.—SATURN AND ITS MOONS.

281. Saturn, which is belted like Jupiter, is surrounded not only by eight moons, but by a series of rings, the innermost one of which is transparent. The belts have been already referred to (Art. 277). Seven of the moons were known for sixty years before the eighth was discovered. Their diameters, distances from their primary, etc., are given in Table III. of the Appendix. The equator of Saturn, unlike that of Jupiter, is greatly inclined to the ecliptic; transits, eclipses, and occultations

---

with our Moon in size? How, in distance from the primary? In what respect do they resemble our Moon? From what is this inferred? 280. What planet ranks next to Jupiter in size? How does Saturn compare with the Earth in size? How do Saturn's day and year compare with ours? 281. How many moons has Saturn? By what else is it surrounded? In what respect is Saturn very unlike Jupiter? In what plane do the orbits of Saturn's satellites mostly

of the satellites, the orbits of which for the most part lie in the plane of the planet's equator and rings, happen but rarely.

282. It is to the rings that most of the interest of this planet attaches. We may imagine how sorely puzzled the earlier observers, with their very imperfect telescopes, were, by these strange appendages. The planet at first was supposed to resemble a vase; hence the name *Ansæ*, or handles, given to the rings in certain positions of the planet. It was next supposed to consist of three bodies, the largest in the middle. The true nature of the rings was discovered by Huyghens in 1655, who announced it in this curious form:—

"aaaaaaa   ccccc   d   eeeee   g   h   iiiiiii   llll   mm   nnnnnnnnn   oooo   pp   q   rr   s   ttttt   uuuuu."

These letters, placed in their proper order, read:—"*Annulo cingitur tenui plano, nusquam cohærente, ad eclipticam inclinato.*"—" It is surrounded by a thin flat ring, nowhere attached to its surface, inclined to the ecliptic."

There is nothing more encouraging in the history of astronomy than the way in which eye and mind have bridged over the tremendous gap that separates us from this planet. By degrees the fact that the appearance was due to a ring was determined; then a separation was noticed, dividing the ring into two; the extreme thinness of the ring came out next, when Sir William Herschel observed the satellites "like pearls strung on a silver thread;" then an American astronomer, Bond, discovered that the number of rings must be multiplied, we know not how many fold. The transparent ring was next made out by Dawes and Bond, in 1852; then the transparent ring

---

lie? What is said of the occurrence of transits, eclipses, and occultations of the satellites? 282. What constitute the most interesting feature of this planet? What was Saturn at first supposed to resemble? What were the rings first called? Who discovered the true nature of the rings, and when? How did Huyghens announce his discovery? State the discoveries successively made re-

148   THE INFERIOR AND SUPERIOR PLANETS.

was discovered to be divided as the whole system had once been thought to be; last of all comes evidence that the smaller divisions in the various rings are subject to change, and that the ring-system itself is probably increasing in breadth, and approaching the planet.

283. Fig. 63 will give an idea of the appearance presented by Saturn and its strange but beautiful appendage; also of its size as compared with the Earth. It will be shown in Chapter XII. that we see sometimes one surface of the ring-system, sometimes another, and occasionally only its edge.

FIG. 63.—SATURN AND THE EARTH—COMPARATIVE SIZE.

284. The ring-system is situated in the plane of the planet's equator, and its dimensions are as follows:—

---

specting the rings of Saturn. 283. Do we always see the same part of the ring-system? 284. In what plane is the ring-system situated? What is the distance

THE RINGS OF SATURN.

	Miles.
Outside diameter of outer ring,	166,920
Inside " "	147,670
Distance from outer to inner ring,	1,680
Outside diameter of inner ring,	144,310
Inside " "	109,100
Inside " dark ring,	91,780
Distance from dark ring to planet,	9,760
Equatorial diameter of planet,	71,900

So that the breadths of the three principal rings, and of the entire system, are as follows:—

	Miles.
Outer bright ring,	9,625
Inner bright ring,	17,605
Dark ring,	8,660
Entire system,	37,570

In spite of this enormous breadth, the thickness of the rings is not supposed to exceed 100 miles.

285. Of what, then, are these rings composed? There is great reason for believing that they are neither solid nor liquid. The idea now generally accepted is that they are composed of myriads of little satellites, moving independently, each in its own orbit, round the planet; giving rise to the appearance of a bright ring when they are closely packed together, and a very dim one when they are most scattered. In this way we may account for the varying brightness of the different parts, and for the haziness on both sides of the ring near the planet (shown in Fig. 84), which is supposed to be due to some of the satellites being drawn out of the ring by the attraction of the planet.

286. Although Saturn appears to resemble Jupiter in its atmospheric conditions, its year, unlike that planet's,

---

from the planet to the nearest dark ring? What is the breadth of the three principal rings respectively? What is the breadth of the entire system? What is the thickness of the rings? 285. Of what are the rings now generally believed to consist? What are accounted for on this supposition? 286. What follows from the great inclination of Saturn's axis, with respect to its seasons? By what be-

150   THE INFERIOR AND SUPERIOR PLANETS.

and like our Earth's, owing to the great inclination of its axis, is sharply divided into seasons. Saturn's seasons, however, are marked by something else than a change of temperature; we refer to the effects produced by the presence of its ring-appendage. To understand these effects, its appearance from the body of the planet must first be considered. As the plane of the ring-system lies in the plane of the planet's equator, an observer at the equator will only see its edge, and the rings will therefore look like a band of light passing through the east and west points and the zenith. As the observer, however, increases his latitude either north or south, the surface of the ring-system will begin to be seen, and will gradually increase in width. As it widens, it will also recede from the zenith, until in lat. 63° it is lost below the horizon; and between this latitude and the poles it is altogether invisible.

Now, the plane of the ring always remains parallel to itself, and twice in Saturn's year—that is, in two opposite points of the planet's orbit—it passes through the Sun. It follows, therefore, that during one-half of the revolution of the planet one surface of the rings is lit up, and during the remaining period the other surface. At night, in the one case, the ring-system will be seen as an illuminated arch, with the shadow of the planet passing over it, like the hour-hand over a dial; and in the other, if it be not lit up by the light reflected from the planet, its position will be indicated only by the entire absence of stars.

287. But if the rings eclipse the stars at night, they can also eclipse the Sun by day. In latitude 40° we have morning and evening eclipses for more than a year, gradually extending until the Sun is eclipsed during the whole

---

sides a change of temperature are its seasons marked? How would the rings be presented to an observer at the equator of Saturn? As he increases his latitude, what changes will be exhibited in the rings? When will the rings sink below the horizon? State the facts with respect to the illumination of the surfaces of the rings. How must the illuminated surface look at night? How is the dark surface indicated? 287. What phenomena are produced by the rings in the day-

URANUS. 151

day—that is, when its apparent path lies entirely in the region covered by the ring. These total eclipses continue for nearly 7 years, and eclipses of one kind or another take place for 8 years 292 days. This will give us an idea how largely the apparent phenomena of the heavens, and the actual conditions as to climates and seasons, are influenced by the presence of the ring.

As the year of Saturn equals $29\frac{1}{2}$ of our years, it follows that each surface of the rings is in turn deprived of the light of the Sun for nearly 15 years.

288. **Uranus** (♅).—Uranus, the next planet to Saturn, usually shines as a star of the sixth magnitude, and is just visible to the naked eye. It is 72 times larger than the Earth, and revolves about the Sun in 84 of our years. There being no spots on its surface, we are unable to fix the period of its revolution on its axis. The intensity of the solar heat and light received at Uranus are only $\frac{1}{310}$ of ours.

Uranus is attended by four moons, which are remarkable for their retrograde motion, travelling in the opposite direction to that of all the other planets, except the moon of Neptune, and in orbits nearly perpendicular to the orbit of their primary. Owing to the great distance of Uranus, nothing is known of its physical peculiarities, except that its specific gravity is about $\frac{1}{4}$ that of the Earth—a little greater than the specific gravity of ice.

289. **Neptune** (♆).—Neptune, the most remote planet of the solar system and the third in size, is invisible to the naked eye. Seen through the telescope, it looks like a star of the eighth magnitude. Its revolution round the Sun is performed in about 165 of our years. It has one

---

time? Describe the eclipses in lat. 40°. How long is each surface of the rings in turn deprived of the Sun's light? 288. What appearance does Uranus present? How do its size, its year, and the intensity of its light and heat compare with ours? How many moons has Uranus? For what are they remarkable? What is the specific gravity of Uranus? 289. Which is the most remote planet of our system? How does Neptune look? How long is its year? How many

moon, a little nearer to its primary than our Moon is to us. Its light and heat have but $\frac{1}{1000}$ the intensity of ours. Nothing is known of the period of rotation or the atmospheric conditions of Neptune. Its density is about $\frac{1}{5}$ of the Earth's, or not quite equal to that of sea-water.

CHAPTER IX.

THE ASTEROIDS, OR MINOR PLANETS.

290. **Bode's Law.**—If we write down

   0   3   6   12   24   48   96

and add 4 to each, we get

   4   7   10   16   28   52   100

and this series of numbers represents very nearly the distances of the ancient planets from the Sun, as follows:—

Mercury, Venus, Earth, Mars, —, Jupiter, Saturn. This singular fact was discovered by Titius, and is known by the name of Bode's Law. We see that the fifth term has apparently no representative among the planets. This fact acted so strongly on the imagination of Kepler that he boldly placed an undiscovered planet in the gap.

291. **Discovery of the Asteroids.**—Up to the time of the discovery of Uranus, the suspected planet had not revealed itself; when it was found, however, that the position of Uranus was very well represented by the next term of Bode's series, 196, it was determined to make an organized search for it. For this purpose a society of astronomers was formed; the zodiac was divided into 24 zones, and each zone was confided to a member of the society. On

---

moons has it? How do Neptune's light, heat, and density, compare with those of the Earth?

290. What is meant by "Bode's Law"? Which term of the series was not represented among the planets? 291. What gave an impetus to the search for

THEIR SIZE AND ORBITS. 153

the first day of the present century a planet was discovered and named Ceres, which, curiously enough, filled up the gap. The discovery of a second, third, and fourth, named respectively Pallas, Juno, and Vesta, soon followed; and up to the present time (Jan., 1883) no less than 231 of these little bodies have been detected. See Table I., Appendix.

292. **Size of the Asteroids.**—None of these planets, except occasionally Ceres and Vesta, can be seen by the naked eye. This is owing to their small size; the largest minor planet is but 228 miles in diameter, and many of the smaller ones are less than 50. The force of gravity on their surfaces must be very small. A man placed on one of them would spring with ease 60 feet high, and sustain no greater shock in his descent than he does on the Earth from leaping a yard. On such planets giants may exist; and those enormous animals which here require the buoyant power of water to counteract their weight, may there inhabit the land.

293. **Orbits.**—The orbits of the asteroids thus far discovered, for the most part, lie nearer to Mars than Jupiter, and are in some cases so elliptical, that, if we take the extreme distances into account, they occupy a zone 240,000,000 miles in width—the distance between Mars and Jupiter being 336,000,000. The planet nearest the Sun is Flora, whose journey round that luminary is performed in 3¼ years, at a mean distance of 201,000,000 miles. The most distant one is Maximiliana, whose year is as long as 6½ of ours, and whose mean distance is 313,000,000 miles.

There is a resemblance between the orbits of some of the asteroids and those of comets, not only in their degree

the undiscovered planet? What plan was adopted? What was the result? What further discoveries have since been made? 292. Which of the asteroids are occasionally visible to the naked eye? What is the size of the asteroids? What follows, with respect to the force of gravity? 293. How do the orbits of the asteroids thus far discovered lie? How wide a zone do they occupy? Which of the asteroids is nearest the Sun? Which is the most distant? How do their mean distances and years compare? In what respects do the orbits of some of the as-

of eccentricity, but also in their great inclination to the plane of the ecliptic. The orbit of Pallas, for instance, is inclined to this plane at an angle of 34°; that of Massilia, on the other hand, nearly coincides with it.

294. **Evidences of Atmosphere and Rotation.**—Pallas has been supposed, from its hazy appearance, to be surrounded by a dense atmosphere, and this may also be the case with the others, as their colors are not the same. There are also evidences that some among them rotate on their axes, like the larger planets.

295. **Mode of Discovery.**—The minor planets lately discovered shine as stars of the tenth or eleventh magnitude; and the only way in which they can be detected, therefore, is to compare the star-charts of different parts of the heavens with the heavens themselves, night after night. Should any point of light not marked on the chart be observed, it is immediately watched, and if any motion is detected, its rate and direction are determined. In the

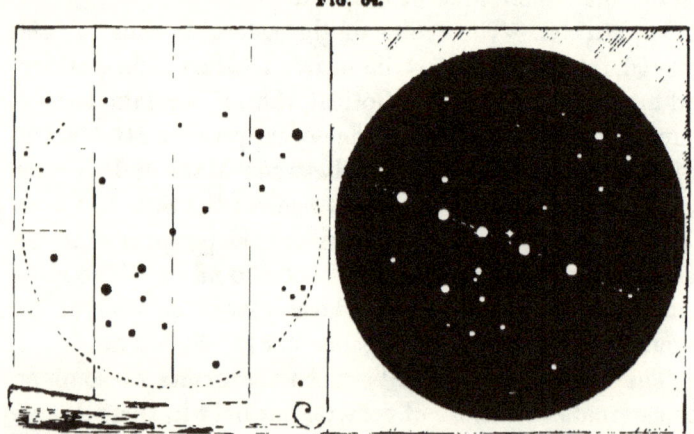

Fig. 64.

latter case, either a new planet or a comet has been discovered.

teroids resemble those of the comets? How do the orbits of Pallas and Massilia differ in inclination to the plane of the ecliptic? 294. Of what are there evidences in the case of some of the asteroids? 295. Describe the mode of discovering the

Fig. 64, which represents on the left the star-map and on the right the field of view of the telescope, will give an idea of the method pursued. The stars shown in both are the same, and in the field of the telescope is seen the new body, which, absent from the map, and changing its position relatively to the stars, is found to be a member of our solar system.

296. **Theory respecting the Asteroids.**—To account for the existence of the asteroids, it has been suggested that they may be fragments of a larger planet destroyed by contact with some other celestial body. One fact seems, above all, to indicate an intimate relation between all the minor planets: it is, that if their orbits are figured under the form of material rings, these rings will be found so entangled that it would be possible, by means of one among them, taken at hazard, to lift up all the rest.

It is probable that the largest of the asteroids have been discovered; yet as more powerful instruments are used, many new ones, less easily visible from the Earth, will no doubt from time to time be added to their number. According to Le Verrier's computation, the total mass of these bodies revolving between the orbits of Mars and Jupiter is such that, allowing them to equal the Earth in density and to have an average size equal to that of such as have been already discovered, their whole number would be not far from 150,000!

CHAPTER X.

COMETS.

297. WE have seen that round the white-hot Sun cold or cooling solid bodies, called planets, revolve; that be-

---

minor planets. 296. What theory has been advanced, to account for the existence of the asteroids? What fact indicates an intimate relation between them? What is probable with regard to future discoveries? How great may the number of the asteroids be?

cause they are cold they do not shine by their own light; that they perform their journeys in almost the same plane; that the shape of their orbits is oval or elliptical; and that they all move in one direction,—that is, from west to east.

But these are not the only bodies which revolve round the Sun. There are, besides, masses, probably white-hot, called **Comets** (from the Greek κομήτης, *long-haired*), which shine by their own light; which perform their journeys round the Sun in every plane, in orbits some of which are so elongated that they can scarcely be called elliptical; and—a further point of difference—while some

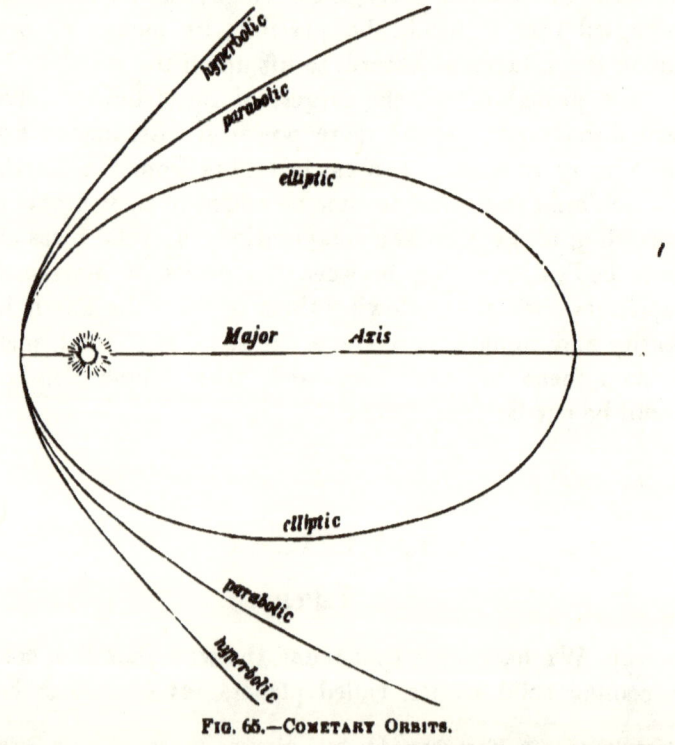

FIG. 65.—COMETARY ORBITS.

297. What have we learned with respect to the planets? What other bodies have we now to consider? How do comets differ from planets? 298. What

move round the Sun in the same direction as the planets, others revolve from east to west.

298. Orbits of the Comets.—The orbits of the comets are either *ellipses*, *parabolas*, or *hyperbolas* (see Fig. 65). Comets that move in elliptical orbits revolve round the Sun in definite periods. Their paths are exceedingly elongated, the cometary orbit which most nearly approaches a circle (that of Faye's comet) having a much greater eccentricity than the planetary orbit which departs most from a circle (the asteroid Polyhymnia's).

Comets that describe parabolas or hyperbolas will never return, as these curves consist of two constantly-diverging branches. After once visiting our system, they go away forever, seeking perhaps another sun in the depths of the heavens. Some of those, however, that appear to describe parabolas may really move in greatly-elongated ellipses and return after very long periods.

Here is a list of some comets whose period is known:—

COMETS.	Time of Revolution.	Nearest Approach to the Sun.	Greatest Distance from the Sun.
	Years.		
Encke's . . .	$3\frac{1}{4}$	32,000,000	387,000,000
De Vico's . .	$5\frac{1}{2}$	110,000,000	475,000,000
Winnecke's .	$5\frac{1}{2}$	—	—
Brorsen's . .	$5\frac{1}{2}$	64,000,000	537,000,000
Biela's . . .	$6\frac{1}{2}$	82,000,000	585,000,000
D'Arrest's . .	$6\frac{1}{2}$	—	—
Faye's . . .	$7\frac{1}{2}$	192,000,000	603,000,000
Mechain's . .	$13\frac{1}{2}$	—	—
Halley's . .	$76\frac{3}{4}$	56,000,000	3,200,000,000

curves do the comets describe? What comets return at fixed periods? How do the elliptical cometary orbits compare in shape with the planetary orbits? What comets will never return? Mention some of the short-period comets, and their time of revolution. Mention some of the long-period comets, and their time of

These are called *Short-period Comets*. Of the *Long-period Comets* we may mention those of 1858, 1811, and 1844, whose periods of revolution have been estimated at 2,100, 3,000, and 100,000 years respectively.

299. **Distances from the Sun.**—From the table given above it will be seen how the distance of these erratic bodies from the Sun varies at different points of their orbits. Thus Encke's comet is twelve times nearer the Sun at perihelion than at aphelion. Some comets whose aphelia lie far beyond the orbit of Neptune, at perihelion almost graze the Sun's surface. Sir Isaac Newton estimated that the comet of 1680 approached so near the Sun that its temperature was two thousand times that of red-hot iron; at its nearest point, it was but one-sixth part of the Sun's diameter from the surface. The comet of 1843 also made a very near approach to the Sun, and was visible in broad daylight.

300. **Appearance presented by a Comet.**—In Fig. 66 we give a representation of Donati's Comet, visible in 1858, which will serve to illustrate a general description of these bodies. The brighter part of the comet is called the *head*, or *coma*; and sometimes the head contains a brighter portion still, called the *nucleus*.

Fig. 66.—Donati's Comet (general view).

The *tail* is the dimmer part flowing from the head; as observed in different comets, it

---

revolution. 299. Give instances showing how the distance of some comets from the Sun varies at different points of their orbits. What facts are stated respecting the comets of 1680 and 1843 ? 300. Name and describe the different parts of a comet. What different appearances does the tail present? What evidence is

may be long or short, straight or curved, single, double, or multiple. The comet of 1744 had six tails, that of 1823 two. In some comets, particularly those whose period of revolution is shortest, the tail is entirely wanting.

Both head and tail are so transparent that all but the faintest stars are easily seen through them. In 1858, the bright star Arcturus was visible through the tail of Donati's comet, at a place where the tail was 90,000 miles in diameter.

301. **Changes in Appearance.**—When these bodies are far away from the Sun, their heat is feeble, and their light dim; we observe them in our telescopes as round misty bodies, moving very slowly, say a few yards in a second, in the depths of space. Gradually, as the comet approaches the Sun, and its motion increases (for the nearer any body, be it planet or comet, gets to the Sun, the faster it travels), the Sun's action begins to be felt, the comet gets hotter, gives out more light, and becomes visible to the naked eye.

A violent action soon commences; the gas bursts forth in jets from the coma toward the Sun, and is instantly driven back again, as the steam of a locomotive going at great speed is driven back on its path, though from a different cause. The jets rapidly change their position and direction, and a tail is formed, which seems to consist of the smoke or products of combustion driven off from the coma, probably by the repulsive power of the Sun, and rendered visible by his light. The tail is always turned away from the Sun, whether the comet be approaching or receding from that body.

As the comet gets still nearer to the Sun, and therefore to the Earth, we begin to see in some instances that the coma contains a nucleus, brighter than itself; the jets

---

there that both head and tail are transparent? Mention a case in point. 301. What changes of appearance take place, as a comet approaches the Sun? In what direction is the tail always turned? What does the coma sometimes con-

are distinctly visible, and sometimes the coma consists of a series of envelopes. This was the case in the beautiful comet of 1858; the nucleus was continually throwing off these envelopes, which surrounded it like the layers of an onion, and peeled off, expanding outward and giving place to others. Seven distinct envelopes were thus seen; as they were driven off, they seemed to be expelled into the tail.

Hence the tails of comets, as a rule, rapidly increase as they approach the Sun, which gives rise to all this violent action.

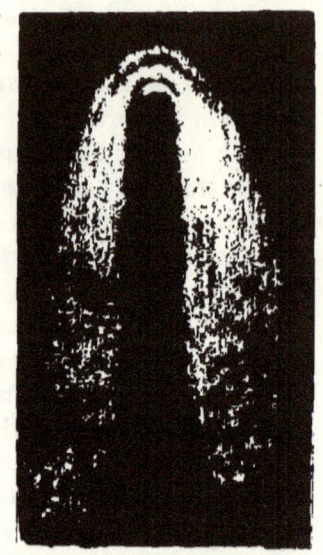

FIG. 67.—DONATI'S COMET (showing the Head and Envelopes).

The tail of the comet of 1861 was 20,000,000 miles long, and this length has been exceeded in many cases. The tail of the comet of 1843 was 112,000,000 miles long, the diameter of the coma being 112,000 miles, that of the nucleus 400 miles; near perihelion, the tail increased at the rate of 35,000,000 miles a day.

Halley's comet, as observed by Sir John Herschel, and Encke's comet, are exceptions to the above rule. As these bodies approached the Sun, both tail and coma decreased, and the whole comet appeared only like a star. Still, most comets increase in brilliancy, and their tails lengthen, as they near the Sun,—so much so that in some instances they have been visible in broad daylight. The enormous effect of a near approach to the Sun may be gathered from the fact that the comet of 1680, at its perihelion passage,

---

tain? What does the nucleus in some cases throw off? What does the tail generally do, as the comet gets still nearer the Sun? What comets were exceptions to this rule? Give the length of the tails of two or three comets. 302. What

while it was travelling at the rate of 1,200,000 miles an hour, in two days shot out a tail 60,000,000 miles long.

302. **Danger from Collisions.**—In old times, when less was known about comets, they caused great alarm; not merely superstitious terror, which connected their coming with the downfall of a king or the outbreak of a plague, but a real fear that they would dash our planet to pieces should they come into contact with it. Modern science teaches us that in the great majority of instances the mass of the comet is so small that we need not be alarmed; indeed, there is good reason to believe that on June 30th, 1861, we actually passed through the tail of the glorious comet which then suddenly appeared, the only noticeable phenomenon being a peculiar phosphorescent mist. Again in 1776, a comet approached so close to Jupiter as to become entangled among its satellites, but the latter all the time pursued their way as if the comet had never existed. This, however, was by no means the case with the comet; it was thrown entirely out of its course, its orbit was changed, and it has ceased to be a long-period comet, its revolution being now accomplished in about twenty years.

303. **A Divided Comet.**—There is an instance on record of a comet's dividing into two portions, which afterward pursued distinct but similar orbits. This is Biela's comet, given in the table in Art. 298. But this is not all. These twin comets were due again at perihelion at the end of January, 1866, and ought to have been visible from the Earth on the 30th of November; but in spite of the strictest watching nothing was seen of them. It is believed that, like Lexell's comet, they have been diverted from their course by some member of our system, and that in this

---

apprehension was formerly felt respecting comets? What does modern science teach us? What two occurrences show us that there is little cause for alarm? 303. What remarkable facts are mentioned in connection with Biela's comet? What is thought to have been the disturbing cause? 304. What have we reason

case the November meteors may have been the disturbing cause.

304. Physical Constitution of Comets.—In the case of a comet without a nucleus, we have reason to believe that the coma is a mass of white-hot gas, like that of which the nebulæ are composed; whether a comet with a nucleus is made up of similar matter we do not know. One thing is certain, that as the tail indicates the waste, so to speak, of the head, each return to the Sun must reduce the mass of the comet.

A diminution of velocity would in time reduce the most refractory comet into a quiet member of the solar family, as the orbit would become less elliptical, or more circular, at each return to perihelion. This effect has, in fact, been observed in some of the short-period comets. Encke's comet, for instance, now performs its revolution round the Sun in three days less than it did eighty years ago. This reduction of speed has been attributed to the resistance offered by the ethereal medium—a resistance not noticed in the case of the planets, because their mass is so much greater.

Sir Isaac Newton calculated that a cubic inch of air at the Earth's surface—that is, as much as is contained in a good-sized pill-box—if reduced to the density of the air 4,000 miles above the surface, would be sufficient to fill a sphere the circumference of which would be as large as the orbit of Neptune. The tail of the largest comet, if it be gas, may therefore weigh but a few ounces or pounds; and the same argument may be applied to the comet itself, if it be not solid. With so limited a supply there is not room for waste, and in the case of so small a mass the resistance offered may easily become noticeable.

---

to believe that a comet consists of? What is certain with respect to each return to the Sun? What has been observed in connection with Encke's comet? To what is this shortening of the period of revolution attributed? What calculation was made by Newton? How much, then, may the tail of the largest comet

305. **Number of Comets.**—From the earliest times, beginning with the Chinese annals, to our own, about 800 comets have been recorded; but the number observed at present is much greater than formerly, as many are revealed by the telescope. It is thought that there may be many millions of these bodies belonging to our system, and perhaps passing between it and other systems. We see but few of them, because those only are visible to us which are favorably situated for observation when they pass the Earth in their journey to or from perihelion; while there may be thousands which at their nearest approach to the Sun are beyond the orbit of Neptune.

306. **Comets, how formerly regarded.**—Comets were in ancient times regarded as the harbingers of war, pestilence, and famine. "A fearful star, for the most part, this comet is," says Pliny, "and not easily appeased." The Sertorian War, the civil dissensions between Pompey and Cæsar, and the cruelties of Nero, were all, as the Romans thought, foreshadowed by these celestial messengers of ill. A comet of terrific sword-like appearance, according to Josephus, seemed to hang over Jerusalem, A. D. 69, inspiring the inhabitants with the greatest alarm; which was realized by the destruction of the city the following year.

As late as the year 1456, all Europe was thrown into consternation by the appearance of a comet, which in the minds of the terrified people presaged victory to the Turks, who were at that time pressing hard upon Christendom. We are told that the bells were rung at noon every day, and prayers offered for preservation "from the Devil, the Turk, and the Comet."

---

weigh, if it be gas? 305. How many comets have been recorded? How does the number now observed compare with that formerly known? Why is this? How many comets, is it thought, may belong to our system? Why do we see so few of them? 306. How were comets regarded in ancient times? What does Pliny say of them? What events in Roman history were thought to have been foreshadowed by comets? By what was the destruction of Jerusalem preceded? What took place as late as the year 1456? What did the Norman Chronicle argue from the comet of 1066 A. D.?

The Norman Chronicle refers to a comet which appeared in the year of the Norman Conquest, 1066, as evidence of the divine right of William of Normandy to invade England.

CHAPTER XI.

METEORS AND METEORITES.

307. **Number of Meteors.**—There are very few clear nights in the year in which, if we watch for some time, we shall not see one of those appearances which are called, according to their brilliancy, Meteors, Bolides, or Shooting-Stars. On some nights we may even see a *shower* of falling stars, and the shower in certain years is so dense that in some places the number seen at once equals the apparent number of the fixed stars seen at a glance. It has been calculated that the average number of meteors which traverse the atmosphere daily, and which are large enough to be visible to the naked eye on a dark clear night, is no less that 7,500,000; and, if we include meteors which would be visible in a telescope, this number will be increased to 400,000,000!

Fig. 68.—Shape of the Zodiacal Light.

Some astronomers have even accounted for the Zodiacal Light, seen at certain seasons, in the east before sunrise, and in the west after sunset, by attributing it to an immense collection of these minute bodies,

307. What are visible on almost any clear night? On some nights what may we even see? How many meteors, on an average, has it been calculated, traverse the atmosphere daily? To what have some astronomers attributed the Zodiacal

which the Earth is thus constantly encountering, and which on entering our atmosphere become meteors. Too small to be seen separately, even with the telescope, they combine in countless multitudes, according to this hypothesis, to reflect to our eyes the light in question, borrowing it from the Sun round which they revolve. The shape of the Zodiacal Light is that of a cone, with its base turned toward the Sun, and its axis nearly in the plane of the ecliptic, as shown in Fig. 68.

308. Theory of the November Showers.—It is now generally held that these little bodies are not scattered uniformly in the space comprised by the Solar System, but are collected into several groups, some of which travel, like comets, in elliptic orbits round the Sun; and that what we call a shower of meteors is due to the Earth's breaking through one of these groups. Two such groups are well defined, and we break through them in August and November in certain years. The exquisitely beautiful star-shower of 1866 has placed the truth of this explanation beyond all doubt.

To explain this theory further, we must again fall back upon our imaginary ocean (Art. 111) to represent the plane of the ecliptic. Let us also suppose that the Earth's path is marked out by buoys placed at every degree of longitude, beginning from the place occupied by the Earth at the autumnal equinox. Now, if it were possible to buoy space in this way, we should see the November group of meteors rising from the plane at the point occupied by our Earth about the 14th of November.

309. But why do we not have a star-shower every November? Because the meteors are not uniformly distributed throughout their orbit, but are mostly collected in a great group in one part of it. To have a dense shower, we must not only cross their orbit, but cross it at a time when the principal group of little bodies is in that part of it

---

Light? What is the shape of this light? 308. What opinion is now generally held respecting meteoric showers? When do we break through two well-defined groups? Illustrate this theory as regards the November group, taking an imaginary ocean to represent the plane of the ecliptic. 309. Why do we not have

which we are crossing. Now, the group referred to performs its revolution in about 33¼ years; hence November showers may be expected at intervals of about 33 years. But as the meteors extend along their orbit in a great stream so far that it takes two or three years for them to clear the node, or point of their path at which they cross the Earth's orbit, a shower, more or less brilliant, proceeding from different parts of the same lengthened group, may occur in two or three consecutive years.

310. About a dozen of these November showers are recorded. One of them occurred November 12th, 1799; another, of remarkable brilliancy, on the 13th of November, 1833, when it was estimated that 575 fell on an average each minute. On the 14th of November, 1866, and also on the same day of the two following years, there was a recurrence of these remarkable phenomena.

311. **The Radiant-point.**—Now, what will happen when the Earth, sailing along in its path, reaches the node and encounters the mass of meteoric dust, the particles of which are travelling in the opposite direction?

Let us in imagination connect the Earth and Sun with a straight line. At any moment, the direction of the Earth's motion will be at right angles to this line (or, as it is called, *a tangent to its orbit*); therefore, as longitudes are reckoned from right to left, the motion will be directed to a point 90° of longitude behind the Sun. The Sun's longitude at noon on the 14th of November, 1866, was 232°, within a few minutes; 90° from this gives us 142°.

Since therefore the meteors, as we encounter them in our journey, should seem to come from the point of space toward which the Earth is travelling, and not from either side, we ought to see them coming from a point situated

---

star-showers every November? How long an interval occurs between successive showers? 310. How many November showers are recorded? Mention three that have taken place. 311. In what direction do the meteoric particles travel, as regards the Earth? If our theory is correct, from what longitude ought the meteors to appear to come? Explain why this is so. Now, what was actually no-

in longitude 142°, or thereabout. Now, what was actually seen?

One of the most striking facts, noticed even by those who did not see its significance, was that all the meteors seen in the star-shower referred to seemed to come from one part of the sky. In fact, there was a region in which the meteors appeared trainless, and shone out for a moment like so many stars, *because they were directly approaching us.* Near this spot they were so numerous, and all so foreshortened and for the most part faint, that the sky at times put on almost a phosphorescent appearance. As the eye travelled from this region, the trains increased in length, those being longest as a rule which first made their appearance overhead, or which trended westward. Now, if the paths of all had been projected backward, they would have intersected in one region, that, namely, in which the most foreshortened ones were seen. In fact, there were moments in which the meteors belted the sky like the meridians on a terrestrial globe, the pole of the globe being represented by a point in the constellation Leo. From this point they all seemed to *radiate*, and *Radiant-point* is the appropriate name given to it by astronomers. Now, the longitude of this point is 142°, or thereabout.

The apparent radiation from this point is an effect of perspective; hence we gather that the paths of the meteors are parallel, or nearly so, and that the meteors themselves all travel in straight lines from the radiant-point.

312. **Orbits of the Meteors.**—By careful observations of the radiant-point it has been determined that the orbit of each member of the November star-shower, and therefore of the whole mass, is an ellipse with its perihelion lying on the Earth's orbit, and its aphelion just beyond the orbit of Uranus; that its inclination to the plane of

---

ticed? What is meant by the Radiant-point? How is it situated? To what is the apparent radiation from this point due? What conclusions do we draw from this? 312. What has been established with respect to the orbit of the me-

the ecliptic is 17°; and that the direction of the motion of the meteors is retrograde.

Up to the present time 56 such radiant-points have been determined, which possibly indicate 56 similar groups moving round the Sun in cometary or planetary orbits. The meteors of particular showers vary in their distinctive characters, some being larger, brighter, and more ruddy, than others—some swifter, and drawing after them more persistent trains, than those of other showers.

313. **Cause of the Luminous Appearance.**—Let us now take the case of a single meteor entering our atmosphere; why does it present such a brilliant appearance?

In the first place, we have the Earth travelling at the rate of 1,100 miles a minute, plunging into a mass of bodies whose velocity, at first equal to its own, is soon increased to 1,800 miles a minute by the Earth's attraction. The meteoric body enters our atmosphere at this rate—30 miles a second. Its motion is soon arrested by the friction of that atmosphere, which *puts a brake on it*, as it were,—and it becomes hot, just as the wheel of a tender gets hot under similar circumstances, or a cannon-ball when the target impedes its further flight. So hot does it get that, at last, as great heat is always accompanied by light, we see it; it becomes vaporized, and leaves a train of luminous vapor behind it.

Heat results from the stoppage of any mechanical force, in proportion to the amount of force stopped. The number of meteors is known to be immense; multitudes beyond conception may exist in the planetary spaces; it has been thought, as already stated, that the Zodiacal Light may even be due to a great ring of these little bodies surrounding the Sun and reflecting its light. Now, putting these facts together, some philosophers have attempted to account for the constant supply of solar heat, kept up without perceptible diminution from century to century.

by attributing it to the stoppage in the solar atmosphere of a succession of meteoric bodies flying toward the Sun with an enormous velocity, accelerated by the solar attraction until they enter this atmosphere, when their force is extinguished by its resistance.

314. **Size and Distance from the Earth.**—All the particles which compose the November shower are small; it has been estimated that some of them weigh but two grains, and that comparatively few exceed a pound. They begin to burn at a height of 74 miles, and are burnt up and disappear at an elevation of 54 miles; the average length of their visible paths being 42 miles. It is supposed that the November-shower meteors are composed of more easily destructible or more inflammable materials than aërolitic bodies.

315. **Other Star-showers.**—What has been said about the appearance of the November meteors applies to the other star-showers, particularly those of August and April, the meteors of which also travel round the Sun in cometary orbits. In fact, there is reason to believe that three bodies which were observed and recorded as comets, were really nothing but meteors, and belonged one to the November, one to the August, and one to the April group.

The August Meteors appear about the 10th of that month, their radiant-point being in the constellation Perseus, and their number inferior to that of the November meteors. It is believed that they are distributed, though in unequal numbers, along their entire orbit, and that this orbit is considerably inclined to the plane of the ecliptic, and extends beyond the orbit of Neptune.

316. **Detonating Meteors.**—In the case of the November and August meteors and shooting-stars generally, the

---

supply of solar heat? 314. What is the size of the particles that compose the November shower? What is their height? Of what kind of materials are they thought to be composed? 315. To what else will what has been said about the November meteors apply? What is there reason to believe respecting three bodies that passed for comets? When do the August meteors appear? Where is their radiant-point? How are they thought to be distributed? What is said of

170   METEORS AND METEORITES.

mass is so small that it is entirely changed into vapor and disappears without noise. There are other classes of meteoric bodies, however, with much more striking effects. At times meteors of unusual brilliancy are heard to explode with great noise; these are called Detonating Meteors. On November 15th, 1859, a meteor of this class passed over New Jersey; it was visible in the full sunlight, and was followed by a series of terrific ex-

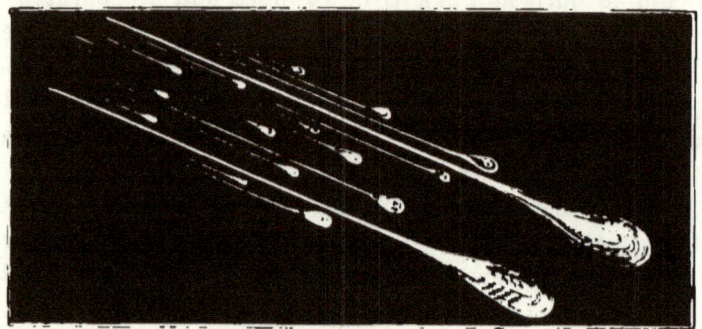

FIG. 60.—FIRE-BALL, AS OBSERVED IN A TELESCOPE.

plosions, which were compared to the discharge of a thousand cannon. Other meteors are so large that they reach the Earth before complete vaporization takes place; and we then have a fall of what are called Meteorites, often accompanied by loud explosions.

317. **Meteorites.**—Meteorites are masses which, owing to their size, resist the action of the atmosphere, and actually complete their fall to the Earth. They are divided into Aërolites, or meteoric *stones ;* Aërosiderites, or meteoric *iron ;* and Aërosiderolites, which comprise the intervening varieties.

318. We do not know whether the meteorites, and meteors which *occasionally* appear, and which are there-

---

their orbit? 316. Why do shooting-stars generally disappear without noise? What other meteoric bodies are there with which this is not the case? Describe a detonating meteor that passed over New Jersey. What happens in the case of very large meteors? 317. What are Meteorites? Into what classes are they di-

fore called Sporadic Meteors, belong to groups or not, although, like the star-showers, they are most common at particular dates. As they are independent of geographical position, it has been thought that there may be some astronomical and perhaps a physical difference between them and the ordinary shooting-stars.

319. **Showers of Aerolites.**—Meteorites of considerable size occasionally reach the Earth's surface, by which men and cattle have been killed, and buildings set on fire. They sometimes fall in showers.

Among the largest aërolitic falls of modern times, we may mention the following. On the 26th of April, 1803, at 2 P. M., a violent explosion was heard at L'Aigle, in Normandy, a luminous meteor having appeared in the air with a very rapid motion a few moments before. Two thousand stones fell, so hot as to burn the hands when touched. The shower extended over an area nine miles long and six miles wide, close to one extremity of which the largest of the stones, weighing nearly twenty-four pounds, was found.

A similar shower of stones fell at Stannern, between Vienna and Prague, on the 22d of May, 1812, when two hundred stones fell on an area eight miles long by four miles wide. The largest stones in this case were found, as before, near the northern extremity of the ellipse. A third stone-fall occurred at Orgueil, in the south of France, on the evening of the 14th of May, 1864. The area over which the stones were scattered was eighteen miles long by five miles wide.

At Kuyahinza, in Hungary, on the 9th of June, 1866, a luminous meteor was seen, and an aërolite weighing six hundredweight, and nearly one thousand smaller stones, fell on an area measuring ten miles in length and four in width. The large mass was found, as in the other cases,

vided? 318. What is meant by Sporadic Meteors? What has been thought respecting them? 319. Describe the fall of aërolites at L'Aigle. At Stannern. At Or-

at one extremity of the oval area. The fall was followed by a loud explosion, and a smoky streak was visible in the sky for nearly half an hour.

320. **Chemical Composition of Meteorites.**—A chemical examination of the fragments of meteorites shows that, although in their composition they are unlike any other natural product, their elements are all known to us, and that they consist of the same materials, though in each variety some particular element may predominate. In the main, they are composed of metallic iron and various compounds of silica, the iron forming as much as 95 per cent. in some cases, and only 1 per cent. in others; hence the division into the three classes referred to in Art. 317. The iron is always associated with a certain quantity of nickel, and sometimes with cobalt, copper, tin, and chromium. Among the silicates may be mentioned augite, and olivine, a mineral found abundantly in volcanic rocks.

Besides these substances, a compound of iron, phosphorus, and nickel, called *schreiberzite*, is generally found; this compound is not a natural terrestrial product, but has been artificially produced. Carbon has also been detected.

321. The chemical elements found in meteorites up to the present time are as follows:—

Metalloids:—Oxygen, sulphur, phosphorus, carbon, silicon.

Metals:—Iron, nickel, chromium, tin, aluminum, magnesium, calcium, potassium, sodium, cobalt, manganese, copper, titanium, lead, lithium, strontium.

322. **Structure of Meteorites.**—The structure of meteorites having been carefully studied with the microscope, it has been ascertained that the matter of which they are composed was certainly at one time in a state of fusion;

---

guell. At Kuyahinza. 320. What is shown by a chemical examination of the fragments of meteorites? Of what are they composed in the main? What are found among the silicates? What is said of *schreiberzite?* 321. Which of the metalloids have been found in meteorites? Which of the metals? 322. What has been ascer

and that the most remote condition of which we have positive evidence was that of small, detached, melted globules. The formation of these cannot be satisfactorily explained except by supposing that their constituents were originally in the state of vapor, as they now exist in the atmosphere of the Sun; and that, on the temperature's becoming lower, they condensed into these "ultimate cosmical particles." These afterward collected into larger masses, which have been variously changed by subsequent metamorphic action, broken up by repeated mutual impact, and often again collected together and solidified. The meteoric irons are probably those portions of the metallic constituents which were separated from the rest by fusion, when the metamorphism was carried to that extreme point.

CHAPTER XII.

APPARENT MOVEMENTS OF THE HEAVENLY BODIES.

323. In the preceding chapters we have studied in detail the whole universe of which we form a part; its nebulæ and stars; the nearest star to us—the Sun; and lastly, the system of bodies which centre in this star, our own Earth being among them.

We should now, therefore, be in a position to see exactly what "the Earth's place in Nature" really is. We find it, in fact, to be a small planet travelling round a small star, and that the whole solar system is but a mere speck in the universe—an atom of sand on the shore, a drop in the infinite ocean of space.

324. **The Earth an Observatory.**—But, however unimportant the Earth may be, compared with the universe generally, or even with the Sun, it is all in all to us who inhabit it, and especially so in an astronomical light; for, although we have in imagination looked at the various celestial orbs from all points of view, our bodily eyes are chained to the Earth. The Earth is, in fact, our observatory, the very centre of the visible creation; and this is why, until men knew better, it was thought to be the very centre of the actual one.

More than this, the Earth is not a *fixed* observatory; it is a *movable* one, and has a double motion, turning round its own axis while it travels round the Sun. Hence, although the stars and the Sun are at rest, they appear to us to move rapidly, and rise and set every twenty-four hours. Though the planets go round the Sun, their circular movements are not visible to us as such, for our own annual movement is mixed up with them.

Having described the heavens, then, as they are, we must describe them as they seem. The *real movements* must now give way to the *apparent ones;* we must, in short, take the motion of our observatory, the Earth, into account.

325. **Apparent Movements, how produced.**—To make this matter clear, let the Earth be supposed to be at rest, neither turning on its axis nor travelling round the Sun. In that case, the side turned toward the Sun would have perpetual day, the other side perpetual night. On the one side, the Sun would appear at rest—there would be no rising and setting; on the other, the stars would be seen at rest in the same manner. The whole heavens would be, as it were, dead.

---

solar system? 324. What makes the Earth important to us, in an astronomical point of view? What kind of an observatory is the Earth? What apparent motions are the result of the Earth's real motions? Why do not the planets appear to move in circles? What movements are we now to consider? 325. If the Earth had no motion at all, what would follow? If the Earth turned on its axis

Again, let us suppose the Earth to go round the Sun as the Moon goes round the Earth, turning once on its axis during each revolution, which would result in the same side of the Earth always being turned toward the Sun. The inhabitants of the illuminated hemisphere would, as before, see the Sun motionless in the heavens; but in this case, those on the other side, although they would never see the Sun, would still see the stars rise and set once a year.

These examples show how the Earth's real movements produce the various apparent movements of the heavenly bodies. The latter are mainly of two kinds—Daily Apparent Movements and Yearly Apparent Movements; the first being due to the Earth's daily rotation on its axis, and the second to the Earth's yearly revolution round the Sun. In each case the apparent movement is, as it were, a reflection of the real one, in the opposite direction; exactly what one observes when travelling smoothly in a railroad train or balloon. When we travel in an express train, the objects appear to fly past us in the opposite direction to that in which we are going; and to the occupants of a balloon, in which not the least sensation of motion is felt, the Earth always seems to fall down from them, or rush up to meet them, when the balloon rises or descends.

The Celestial Sphere.

326. **The Celestial Sphere.**—We shall first study the effects of the Earth's rotation on the apparent movements of the stars.

The daily motion of the Earth is very different in different parts—at the equator and poles, for instance. An

---

but once during each revolution round the Sun, what would be the result? What should these examples show us? Into what classes may the apparent movements be divided? Of what is the apparent movement in each case a reflection? Illustrate this. 326. How does the daily motion of the Earth differ in different parts? What should we expect to see in consequence? What do we see? What

observer at a pole is simply turned round without changing his place, while one at the equator is swung round a distance of nearly 25,000 miles every day. We ought, therefore, to expect to see corresponding differences in the apparent motions of the heavens, if they are really due to the actual motions of our planet. Now, this is exactly what is observed. Not only is the apparent motion of the heavens from east to west—the real motion of the Earth being from west to east—but those parts of the heavens which are over the poles appear at rest, while those over the equator appear in most rapid motion. In short, the apparent motion of the Celestial Sphere—the name given to the visible vault of the sky—to which the stars appear to be fixed, and to which their positions are always referred, is exactly similar to the real motion of the terrestrial sphere, our Earth; but, as we said before, in an opposite direction.

Before proceeding further, however, we must explain the terms applied to different parts of the Celestial Sphere.

327. **Celestial Poles and Equator; Zenith and Nadir.**—In the first place, as the stars are so far off, we may imagine the centre of the Celestial Sphere to lie either at the centre of the Earth or in our eye, and we may imagine it as large or as small as we please. The points at which the Earth's axis would pierce this sphere, if it were extended at each end, we call the Celestial Poles; the great circle lying in the same plane as the terrestrial equator is distinguished as the Celestial Equator, or Equinoctial. The point overhead is the Zenith; the point beneath our feet, the Nadir.

328. **Declination and Right Ascension.**—As the Earth is belted by *parallels of latitude* and *meridians of longi-*

---

is the Celestial Sphere? To what is the apparent motion of the celestial sphere similar? 327. Where may we imagine the centre of the celestial sphere to lie? What is meant by the Celestial Poles? What is the Celestial Equator? The Zenith? The Nadir? 328. By what are the heavens belted, to the astronomer?

DECLINATION AND RIGHT ASCENSION. 177

tude, so are the heavens belted to the astronomer with *parallels of declination* and *meridians of right ascension.*

If we suppose the plane in which our equator lies, extended to the stars, it will pass through all the points which have no declination (0°). Above and below this plane we have north and south declination, as we have north and south latitude, till we reach the pole of the equator (90°).

As we start from the *meridian of Greenwich* in reckoning *longitude*, so do we start from a certain point in the celestial equator occupied by the Sun at the vernal equinox, called the *first point of Aries*, in measuring *right ascension*. As we say such a place is so many degrees east of Greenwich, so we say such a star is so many hours, minutes, or seconds, east of the first point of Aries.

In short, as we define the position of a place on the Earth by saying that its latitude and longitude (in degrees) are so-and-so, in like manner do we define the position of a heavenly body by saying that, referred to the celestial sphere, its declination (in degrees) and right ascension (in time reckoned from Aries) are so-and-so.

329. Sometimes the distance from the north celestial pole is given instead of that from the celestial equator. This is called North-polar Distance, and is denoted in almanacs, etc., by the initials N. P. D. As the pole is 90° from the equator, the north-polar distance is evidently the *complement* of the declination—that is, the difference between the declination and 90°.

330. **The Horizon.—Altitude, Azimuth, etc.**—The terms defined above apply to the celestial sphere generally. When we consider that portion of it visible in any one

---

To what on the Earth do these lines correspond? What in the heavens correspond to latitude and longitude on the Earth? From what is declination measured? From what, right ascension? Illustrate the way in which these terms are used. 829. To locate a point in the heavens, what is sometimes given instead of its declination? What relation do the north-polar distance and the declination bear to each other? 330. What is meant by the Visible or Sensible Horizon?

12

place, or the *sphere of observation*, there are other terms employed, which we proceed to explain.

In any place the visible portion of the celestial sphere seems to rest on either the Earth or the sea. The line where the heavens and Earth seem to meet is called the Visible or Sensible Horizon; the Plane of the Visible Horizon meets the Earth at that point of the surface which is occupied by the observer. The Rational, or True Horizon, is a great circle of the heavens, the plane of which is parallel to the former plane, but which, instead of being a tangent to the Earth's surface, passes through its centre.

331. **A Vertical Line** is a line passing from the zenith to the nadir, and therefore through the observer; it is clearly at right angles to the planes of the horizon.

332. If it is desired to point out the position of a heavenly body, not on the celestial sphere generally, but on that portion of it visible above the horizon of a place at a given moment, this is done by determining either its *altitude* or *zenith-distance*, and its *azimuth* (instead of its declination and right ascension).

Altitude is the angular height above the horizon.

Zenith-distance is the angular distance from the zenith. As the zenith is 90° from the horizon, the zenith-distance is evidently the complement of the altitude—that is, the difference between the altitude and 90°.

Azimuth is the angular distance between two planes, one of which passes through the north or south point (according to the hemisphere in which the observation is made), and the other through the object, both passing through the zenith. Azimuth is to Altitude what longitude is to latitude, or what right ascension is to declination.

333. **The Celestial Meridian** of any place is the great circle on the sphere corresponding to the terrestrial meridian of that place, cutting therefore the north and south points.

The Prime Vertical is another great circle passing through the east and west points and the zenith.

Apparent Movements of the Stars.

334. **Rising, Culmination, and Setting of the Stars.**—We are now prepared to proceed with our inquiry into the apparent movements of the celestial sphere. We shall continue to speak of the Sun or a star as rising or setting, although the student now understands that it is, in fact, the plane of the observer's horizon which changes its direction with regard to the heavenly body, in consequence of its being carried round by the Earth's motion. When a star is so situated that it is just visible on the eastern horizon, it is said to *rise*. When the rotation of the Earth has brought the plane of the horizon under the meridian which passes through the star, the latter is said to *culminate* or pass the meridian. When the plane of the horizon is carried to the nadir of the point it passed through when the star first became visible, the star appears on the opposite—that is, the western—horizon, and is said to *set*.

335. **Apparent Movements, as seen from Different Parts of the Earth.**—Let Fig. 70 represent the celestial sphere, and N an observer at the north pole of the Earth. To him the north pole of the heavens (P) and the zenith (Z) coincide, and his true horizon is the celestial equa-

---

by the Celestial Meridian of any place? What is the Prime Vertical? 834. When we speak of the Sun or a star as rising or setting, what do we really mean? When is a star said to *culminate?* 335. With Fig. 70, explain the apparent movements

tor. Above his head is the pivot on which the heavens appear to revolve, as beneath his feet is the pivot on which the Earth actually revolves; and round this point the stars appear to move in circles, which get larger and larger as the horizon is approached. The stars never rise or set, but always keep the same distance from the horizon. The observer is merely carried round by the Earth's rotation, and the stars seem to be carried round in the opposite direction.

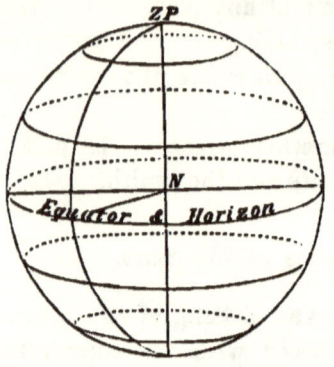

FIG. 70.—CELESTIAL SPHERE, VIEWED FROM THE POLES. A PARALLEL SPHERE.

336. We will now change our position. In Fig. 71 the celestial sphere is again represented, but this time we suppose an observer, Q, at its centre, to be on the Earth's equator. In this position we find the celestial equator in the zenith, and the celestial poles PP on the true horizon, and the stars, instead of revolving round a fixed point overhead, and never rising or setting, rise and set every twelve hours, travelling straight up and down along circles which get smaller and smaller as we leave the zenith and approach the poles. The spectator is carried up and down by the Earth's rotation, and the stars appear to be so carried.

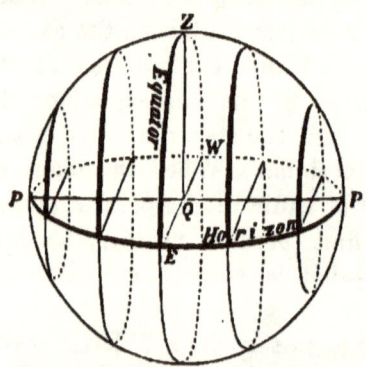

FIG. 71.—THE CELESTIAL SPHERE, VIEWED FROM THE EQUATOR. A RIGHT SPHERE.

of the stars, as seen from the north pole. 336. How is the observer supposed to be placed in Fig. 71? In this position, where do we find the celestial equator, and where the poles? How do the stars appear to move, and why? 337. In Fig.

AN OBLIQUE SPHERE. 181

337. Yet another figure, to show what happens half-way between the poles and the equator. At O, in Fig. 72, we imagine an observer to be placed on our Earth in latitude 45° (that is, half-way between the equator in lat. 0°, and the north pole in lat. 90°). Here the north celestial pole will be half-way between the zenith and the horizon (see Figs. 70 and 71); and close to the pole he will see the stars describing circles, inclined, however, and not retaining the same distance from the horizon. As the eye leaves the pole, the stars rise and set obliquely, and describe larger circles, gradually dipping more and more under the horizon, until, when the celestial equator, $B B' B''$, is reached, half their journey is performed below it. Farther south, we find the stars rising less and less above the horizon; and finally, as there were northern stars that never dip below the horizon, so there are southern stars which never appear above it.

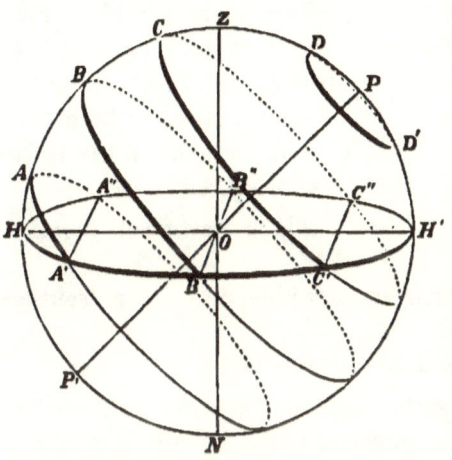

FIG. 72.—CELESTIAL SPHERE, VIEWED FROM A MIDDLE LATITUDE. AN OBLIQUE SPHERE. $D D'$ represents the apparent path of a circumpolar star; $B B' B''$, the path and rising and setting points of an equatorial star; $C C' C''$ and $A A' A''$, those of stars of mid-declination, one north and the other south.

To observers in lat. 45° south, the southern celestial pole is in like manner visible; the stars we never see in the northern hemisphere, never set; the stars which never set with us, never rise; the stars which rise and set with us, set and rise with them.

72, where do we suppose the observer to be placed? Explain the movements of the stars, as seen from this point. What appearances would be presented in lat.

338. **Stars visible in Different Latitudes.**—Now let the celestial sphere be divided into two hemispheres, a northern and a southern: it is evident that an observer at the north pole sees only the stars of the northern hemisphere; one at the south pole, only those of the southern; while one at the equator sees both. An observer in a middle north latitude sees all the northern stars and some of the southern ones; and another in a middle southern latitude sees all the southern stars and some of the northern ones.

Hence, in middle latitudes, and therefore in the United States, we may divide the stars into three classes:—

I. Those northern stars which never set (northern circumpolar stars).
II. Those southern stars which never rise (southern circumpolar stars).
III. Those stars which both rise and set.

339. It is easily gathered from Figs. 70, 71, 72, that the height of the celestial pole above the horizon at any place is equal to the latitude of that place; for at the equator, in lat. 0°, the pole was on the horizon, and consequently its altitude was nothing; at the pole, in lat. 90°, it was in the zenith, and its altitude was therefore 90°; while in lat. 45° its altitude was 45°. Accordingly, at New York, in lat. 40¾°, its altitude will be 40¾°; hence, stars of less than that distance from the pole will always be visible, as they will be above the horizon when passing below the pole. All the stars, therefore, within 40¾° of the north pole will form Class I.; all those within 40¾° of the south pole, Class II.; and the remainder—that is, all stars from 49¼° N. (90°−40¾°=49¼°) to 49¼ S. will form Class III.

<hr>

45° south? 338. What stars are visible to an observer at the north pole? At the south pole? At the equator? In a middle northern latitude? In a middle southern latitude? In middle latitudes, how may the stars be divided? 339. What is the height of the celestial pole above the horizon at any place? Illustrate this in the case of New York, and state what stars will there belong to each of the three classes just specified. What stars belong to each of the three classes, in the latitude in which you live? 340. What may be used with advan-

340. **Use of the Globes.**—In these and similar inquiries the use of the terrestrial and the celestial globe is of great importance. To use either properly, we must begin by making each a counterpart of what is represented; that is, the north pole must be north, the south pole south, and the axis of either globe must be made parallel to the Earth's axis. To find the north point, a compass may be used, allowance being made for its variation from the true meridian, as established for the time and place in question.

The brazen meridian being thus made to run due north and south, the pole—the north pole in our case—must be elevated to correspond with the latitude of the place where the globe is used. At the poles this would be 90°, at the equator 0°, and at New York 40¾°. The wooden horizon, if perfectly level, will then represent the true horizon of the place.

If we now turn the terrestrial globe from west to east, we exactly represent the Earth's motion; and, turning the celestial globe from east to west, we have an exact representation of the apparent movements of the stars for the place in question. It will be seen that some stars never descend below the true horizon, while others never rise above it.

341. **Position of the North Celestial Pole.**—At the present time the north celestial pole lies in Ursa Minor, and a star in that constellation very nearly marks the position of the pole, and is therefore called Polaris, or the Pole-star. The direction in which the Earth's axis points is not always the same, although it varies so slowly that a few years do not make much difference. As a consequence, the position of the celestial poles, which are the

---

tage in these Inquiries? How are the globes prepared for use? How may the Earth's motion and the apparent movements of the stars be exactly represented? 341. Where is the north celestial pole at present situated? What star marks its position? What causes change in the position of the celestial poles? 342. What

points where the Earth's axis prolonged would strike the celestial sphere, varies also.

342. **The Circumpolar Constellations.**—One of the most striking northern circumpolar constellations is Ursa Major or the Great Bear (see Fig. 17, p. 30), also called Charles's Wain and the Plough. Seven bright stars in this constellation (connected by lines in Fig. 17) form what is called the Great Dipper, three making the handle and four the bowl. The two stars of the bowl which are farthest from the handle (Merak and Dubhe) are called the Pointers, because the straight line which connects them points very nearly to the north pole, in whatever position the constellation may be. This will be seen from Fig. 74.

The Pole-star, readily found with the aid of the Pointers, is at the extremity of the tail of the Little Bear, and unites with six other stars of that constellation to form

FIG. 73.—THE NORTHERN CIRCUMPOLAR CONSTELLATIONS.

the Little Dipper (see Fig. 73). This resembles the

---

is one of the most striking of the northern circumpolar constellations? How is the Great Dipper formed? What stars are called the Pointers, and why? In what constellation is the Pole-star? What does it help to form? Name the other more important northern circumpolar constellations. What is Cassiopea sometimes called, and why? 343. What are the principal

Great Dipper in shape, but is smaller and formed of less brilliant stars.

The other more important northern circumpolar constellations are Cassiopea, Cepheus, Camelopardalus, and Draco. Seven stars in Cassiopea form what may be fancied to be the outline of a chair (see Fig. 73), and hence this constellation is sometimes called the Lady's Chair.

343. The principal southern circumpolar constellations are Crux (the Cross), Triangulum Australe (the Southern Triangle), Ara (the Altar), Pavo (the Peacock), Toucanus (the Toucan), Hydrus (the Water-snake), and Dorado (the Sword-fish). All these can be found on the Celestial Chart of the Southern Hemisphere. The other constellations mentioned in Arts. 55, 56, and 57, belong to Class III., both rising and setting to observers in the United States.

344. **Period of the Apparent Movements of the Celestial Sphere.**—As the Earth's rotation is accomplished in 23h. 56m. 4s., it follows that the apparent movement of the celestial sphere is completed in that time; and were there no clouds, and no Sun to eclipse the stars in the daytime by his superior brightness, we

FIG. 74.—DIFFERENT POSITIONS OF THE GREAT DIPPER AT INTERVALS OF SIX HOURS.

---

southern circumpolar constellations? To what class do the other constellations belong? 344. In what time is the apparent movement of the celestial sphere completed, and why? If there were no clouds, what should we

should see the grand procession of distant worlds ever defiling before us, and commencing afresh after that period. The circumpolar constellations would be seen to make a complete revolution round the Pole-star. The Great Dipper, for instance, would at intervals of six hours occupy the different positions shown in Fig. 74.

345. **Effect of the Earth's Yearly Revolution on the Apparent Movements of the Stars.**—We see stars only at night, because in the daytime the Sun puts them out; and the particular stars we see on any given night are those which occupy that half of the celestial sphere opposite to the Sun. Now, as we go round the Sun, we are at different times on different sides, so to speak, of the Sun; and if we could see the stars beyond him, we should see them change. But what we cannot do at mid-day, in consequence of the Sun's brightness, we can easily do at midnight; for, if the stars behind the Sun change, the stars exactly opposite to his apparent place will change too, and these we can see in the south at midnight.

It is clear, in fact, that in one complete revolution of the Earth round the Sun every portion of the visible celestial sphere will in turn be exposed to view in the south at midnight. As the revolution is completed in 365 days, and there are 360° in a great circle of the sphere, we may say that the portion of the heavens visible in the south at midnight advances about 1° from night to night. This 1° is passed over in 4 minutes, as the whole 360° are passed over in nearly 24 hours.

346. This advance is a consequence of the difference between the lengths of the day as measured by the fixed stars and by the moving Sun. As the solar day is longer than the sidereal day, the stars by which the latter is

---

see? What would the circumpolar constellations be seen to do? 345. What other change is seen in the heavens, and to what is it due? At what rate does the portion of the heavens visible in the south at midnight advance, and why? 346. Of what is this advance a consequence? How do the solar and the sidereal

measured gain upon the former at the rate we have mentioned; so that, as seen at the same hour on successive nights, the whole celestial vault advances to the westward, the change due to one month's apparent yearly motion being equal to that brought about in two hours by the apparent daily motion.

Hence the stars south at midnight (or opposite the Sun's place) on any night, were south at 2 A. M. a month previously, and so on; and will be south a month hence at 10 o'clock P. M., and so on.

347. **How to Identify the Stars in the Sky.**—A knowledge of the various stars and constellations may be obtained with the aid of a celestial globe.* We first, as already stated, place its brass meridian in the plane of the meridian of the place in which the globe is used, and make the axis of the globe parallel to the axis of the Earth and the heavens, by elevating the north pole until its height above the wooden horizon is equal to the latitude of the place. We next bring under the brazen meridian the actual place in the heavens occupied by the Sun at the time; this place is given for every day in the almanacs. We thus represent exactly the position of the heavens at mid-day, and the index is then set at 12; for it is always 12, or noon, at a place, when the Sun is in the meridian of that place. Then, if the time at the place is

* WHITALL'S MOVABLE PLANISPHERE, widely known and used in this country, is undoubtedly superior to any celestial globe or map for imparting an accurate acquaintance with the stars. This ingenious instrument, set for a given day of the month and a given hour and minute, held with the zenith overhead, and the extremity of the meridian marked north toward the north, shows the constellations and principal stars visible at the time (and only these) in the exact positions which they then occupy in the heavens, so that they can be distinguished and named with the utmost ease. A variety of problems may be solved with the Planisphere, the use of which invests the study with a practical interest which is truly surprising.—*American Editor.*

day compare in length? In how long a time will the daily motion produce changes equal to those produced by one month's yearly motion? What follows with respect to the stars south at midnight on any given night? 347. How may a knowledge of the stars be obtained? How is the globe rectified? When it is

after noon, we turn the globe from east to west—if before noon, from west to east—till the index and the time correspond.

When the globe has thus been *rectified*, as it is called, we have the constellations which are rising on the eastern horizon, just appearing above the eastern part of the wooden horizon. Those setting are similarly near the western part of the wooden horizon. The constellations in the zenith at the time will occupy the highest part of the globe, while the constellations actually on the meridian will be underneath the brazen meridian.

348. Further, it is easy at once to see at what time any stars will rise, culminate, or set, when the globe is rectified in this manner. All that is necessary is, as before, to bring the Sun's place, given in the almanac, to the meridian, and set the index to 12. To find the time at which any star rises, we bring it to the eastern edge of the wooden horizon, and note the time, which is the time of rising. To find the time at which any star sets, we bring it similarly to the western edge of the wooden horizon and note the time, which is the time of setting. To find the time at which any star culminates, we bring the star under the brazen meridian and note the time, which is the time of meridian passage.

349. In the absence of the celestial globe [or planisphere], the student will derive assistance from the following table, which indicates the positions occupied by the constellations at certain hours during each month in the year. The constellations should be looked for in the sky in the positions specified, and compared with the Celestial Chart at the end of the volume, where the names of the bright stars they contain will be found. The stars in

---

thus rectified, where are the constellations rising, setting, and at the zenith, respectively seen? 348. How can it be found when a given star will rise, set, or culminate? 349. What is furnished, to aid the student in finding the stars and constellations, in the absence of the celestial globe or planisphere? What suggestions are made as to the use of the table?

their vicinity may then be traced. A little study and a few comparisons of the heavens with the charts will soon familiarize the pupil with the principal stars and constellations, and their relative positions.

CONSTELLATIONS VISIBLE IN THE UNITED STATES ON DIFFERENT EVENINGS THROUGHOUT THE YEAR.*

JAN. 20, 10 P. M.
(Feb. 4, 9 P. M.; Feb. 19, 8 P. M.; Dec. 21, midn't; Jan. 5, 11 P. M.)
N—S. Draco, *polaris*, Camelopardalus, *Auriga, Orion, Lepus, Columba Noachi.
E—W. Leo, Cancer, * Aries, Cetus.
NE—SW. Canes Venatici, Ursa Major, Lynx, * Taurus, Eridanus.
SE—NW. Monoceros, Gemini, * Perseus, Andromeda.

Look for Capella a little west or north-west of the zenith; and Betelgeuse, or *a Orionis*, very near the meridian, about one-third of the way from the zenith to the southern horizon. A little west of Betelgeuse is Bellatrix.

FEB. 20, 9½ P. M.
(Feb. 27, 9 P. M.; Mar. 15, 8 P. M.; Jan. 13, midn't; Jan. 28, 11 P. M.)
N—S. Draco, Ursa Minor, *polaris*, *Lynx, Gemini, Canis Minor, Monoceros, Ship Argo.
E—W. Virgo, Leo Minor, * Auriga, Taurus.
NE—SW. Boötes, Canes Venatici, Ursa Major, * Orion, Eridanus.
SE—NW. Hydra, Leo, * Perseus, Andromeda.

Look for Castor very near the zenith, a little to the west. Near it is Pollux, on the meridian, and in the zenith in lat. 31°. Procyon, in Canis Minor, is very nearly on the meridian, about one-third of the distance from the zenith to the southern horizon. The Milky Way will be seen running along the heavens, in a curve west of the meridian and not far distant from it, from the northern to the southern horizon.

MARCH 21, 10 P. M.
(Apr. 5, 9 P. M.; Apr. 20, 8 P. M.; Feb. 19, midn't; Mar. 6, 11 P. M.)

* The asterisk on the several lines denotes that the zenith (in lat. 40°, to which the table particularly refers, though it will serve for any lat. in the U. S.) separates the two constellations between which the asterisk is placed. When the asterisk is prefixed to any constellation, the constellation itself occupies the zenith.

190   CONSTELLATIONS VISIBLE IN THE U. S.

N—S.   Cepheus, *polaris*, Ursa Major, * Leo Minor, Leo, Hydra.
E—W.   Virgo, Coma Berenices, * Gemini, Orion.
NE—SW. Hercules, Corona Borealis, Boötes, Canes Venatici, *
       Cancer, Monoceros, Canis Major.
SE—NW. Hydra, Virgo, * Lynx, Camelopardalus, Perseus.

Look for Regulus, the brightest star of the Lion, on the meridian, nearly one-third of the way from the zenith to the southern horizon.

APRIL 20, 10 P. M.
(May 5, 9 P. M.; May 20, 8 P. M.; Mar. 20, midn't; Apr. 5, 11 P. M.)

N—S.   Cassiopea, Cepheus, *polaris*, Ursa Major, * Coma Berenices, Virgo, Corvus.
E—W.   Ophiuchus, Hercules, Corona Borealis, Boötes, Cor Caroli, * Leo Minor, Cancer, Canis Minor, Monoceros.
NE—SW. Lyra, Draco, * Leo Minor, Leo, *a Hydræ*.
SE—NW. Libra, Boötes, * Lynx, Auriga.

Look for Denebola, or β *Leonis*, a short distance west of the meridian, and a little less than ⅓ of the way from the zenith to the southern horizon. *Alpha* (a) *Hydræ*, in the south-west, otherwise called *Cor Hydræ* and Alphard, is a variable star, ranging from the second to the third magnitude. It can be easily found from the fact that there is no other large star near it.

MAY 21, 10 P. M.
(June 5, 9 P. M.; May 28, 9½ P. M.; May 6, 11 P. M.; Apr. 20, midn't.)

N—S.   Cassiopea, *polaris*, η *Ursæ Majoris* or *Ackair*, * Arcturus or *a Boötis*, Virgo, Centaurus.
E—W.   Aquila, Hercules, * Canes Venatici, Cor Caroli, Leo Minor, Leo, *a Hydræ*.
NE—SW. Vulpecula et Anser, Lyra, Hercules, * Coma Berenices, Virgo, Crater.
SE—NW. Scorpio, Ophiuchus, Serpens, Boötes, * Ursa Major, Lynx, Gemini.

Look for Spica, or *a Virginis*, west of the meridian, and more than half-way from the zenith to the southern horizon.

JUNE 21, 10 P. M.
(July 6, 9 P. M.; June 29, 9½ P. M.; June 5, 11 P. M.; May 22, midn't.)

N—S.   Camelopardalus, *polaris*, Ursa Minor, Draco, * Hercules, Serpens, Scorpio.

ON DIFFERENT EVENINGS OF THE YEAR. 191

E—W. Delphinus, Vulpecula et Anser, Lyra, ٭ Boötes, Canes
Venatici, Coma Berenices, Leo.
NE—SW. Pegasus, Cygnus, ٭ Boötes, Virgo.
SE—NW. Sagittarius, *Hercules, Ursa Major, Lynx.

JULY 22, 10 P. M.

(Aug. 5, 9 P. M.; July 30, 9¼ P. M.; July 7, 11 P.M.; June 22, midn't.)
N—S. Camelopardalus, *polaris*, Draco, ٭ Lyra, Sagittarius.
E—W. Pegasus, Cygnus, ٭ Boötes, Virgo.
NE—SW. Andromeda, Cepheus, ٭ Hercules, Serpens, Libra.
SE—NW. Capricornus, Delphinus, Vulpecula et Anser, Lyra,
٭ Ursa Major, Leo Minor.

Look for the bright star *a Lyræ*, otherwise called Vega and Lyra, very near the zenith, a little east of the meridian. Northeast of Vega, and very near it, is *ε Lyræ*, the remarkable double-double star described in Art. 66.

AUG. 23, 10 P. M.

(Sept. 7, 9 P. M.; Aug. 31, 9¼ P. M.; Aug. 8, 11 P. M.; July 24, midn't.)
N—S. Camelopardalus, *polaris*, Cepheus, Draco, *Cygnus,
Vulp. et Anser, Delphinus, Antinous, Capricornus.
E—W. Pisces, *a Andromedæ* or *Alpheratz*, ٭ Lyra, Hercules,
Serpens, Libra.
NE—SW. Perseus, Andromeda, *Cygnus, Ophiuchus.
SE—NW. Aquarius, Pegasus, ٭ Hercules, Boötes, Canes Venatici.

Nearly midway between the zenith and the eastern horizon, look for the Square of Pegasus, formed by four double stars of the second magnitude, Scheat and Markab at the western angles, and Alpheratz (*a Andromedæ*) and Algenib (*γ Pegasi*) at the eastern. The two stars last named are on the First Meridian, and with Caph (β *Cassiopeæ*), which is on the same line 30° north of Alpheratz, and Polaris, serve to define the position of the great circle from which right ascension is measured.

SEPT. 23, 10 P. M.

(Oct. 16, 8¼ P. M.; Oct. 1, 9¼ P. M.; Sept. 7, 11 P. M.; Aug. 24, midn't.)
N—S. Ursa Major, *polaris*, Cepheus, ٭ Pegasus, Aquarius,
Piscis Australis.
E—W. Aries, Andromeda, ٭ Cygnus, Ophiuchus.
NE—SW. Auriga, Cassiopea, ٭ Delphinus, Antinous, Sagittarius.
SE—NW. Cetus, Pisces, *a Andromedæ*, ٭ Cygnus, Hercules.

Oct. 23, 10 p. m.
(Nov. 7, 9 p. m.; Nov. 23, 8 p. m.; Sept. 23, midn't; Oct. 8, 11 p.m.)
N—S. Ursa Major, *polaris*, Cassiopea, * Andromeda, Pisces, Cetus.
E—W. Orion, Taurus, Perseus, * Pegasus, Delphinus, Antinous.
NE—SW. Gemini, Auriga, Perseus, * Pegasus, Aquarius, Capricornus.
SE—NW. Eridanus, Cetus, Aries, * Andromeda, Cygnus, Lyra, Hercules.

The meridian now very nearly corresponds with the First Meridian. Caph, or β *Cassiopeæ*, will be found nearly due north of the zenith, and Alpheratz and Algenib a few degrees south of the zenith. Fomalhaut, a first-magnitude star in Piscis Australis, is on the meridian at 38 minutes past 8, Oct. 23, and may be seen near the southern horizon.

Nov. 22, 10 p. m.
(Dec. 7, 9 p. m.; Dec. 23, 8 p. m.; Oct. 23, midn't; Nov. 7, 11 p. m.)
N—S. η *Ursæ Majoris*, Draco, *polaris*, * Perseus, Triangulum, Aries, Cetus.
E—W. Monoceros, Auriga, * Andromeda, Pegasus, Aquarius.
NE—SW. Cancer, Lynx, * Pisces, Aquarius, Piscis Australis.
SE—NW. Lepus, Orion, Taurus, * Andromeda, Cygnus.

Look for the triple star Almaack, or γ *Andromedæ*, very near the zenith; in its neighborhood is an elliptical nebula, described in Art. 95. A little west of Almaack is a nebula of minute stars visible to the naked eye, supposed to be the nearest of all the great nebulæ. The variable star Mira, described in Art. 74, is now almost on the meridian, about half-way between the zenith and the southern horizon. The interesting star Algol, or β *Persei* (Art. 75), will be on the meridian, and at the zenith in lat. 40°, at 47 minutes past 10, Nov. 22.

Dec. 21, 10 p. m.
(Jan. 5, 9 p. m.; Jan. 20, 8 p. m.; Nov. 21, midn't; Dec. 6, 11 p. m.)
N--S. Draco, Ursa Minor, *polaris*, Camelopardalus, * Perseus, Taurus, Eridanus.
E—W. Hydra, Cancer, Gemini, Auriga, * Triangulum, Pisces.
NE—SW. Leo Minor, Ursa Major, Lynx, * Aries, Cetus.
SE—NW. Canis Major, Monoceros, * Andromeda, Pegasus.

About 15° south of the zenith, and a little west of the meridian, is the group of the Pleiades (Art. 86), consisting of six stars visible to the naked eye, the brightest of which is Alcyone, of the third magnitude. South-east of the Pleiades 11°, and just east of the meridian, are the Hyades, a group which may readily be recognised by the brilliancy of its principal star, Aldebaran, or *a Tauri*. Aldebaran reaches the meridian at 25 minutes past 10, Dec. 21, or at 9 o'clock on Jan. 11.

FIG. 75. EQUATORIAL CONSTELLATIONS, VISIBLE IN THE SOUTH ON JAN. 20, AT 10 P. M.

350. In Fig. 75 are shown some of the equatorial constellations visible in the south on the 20th of January. The central one is Orion, one of the most marked in the heavens. When all the bright stars in Orion are known, many of the surrounding ones may easily be found, by means of alignments. For instance, the line formed by the three stars in the belt, if produced in a south-easterly direction, will pass near Sirius, the brightest star in the heavens, and prolonged in the opposite direction will nearly pass through Aldebaran. Sirius, Betelgeuse, and

350. What does Fig. 75 show? What is said of Orion? How may Sirius and Aldebaran be found? What three bright stars form a triangle in this part of

Procyon, form a triangle whose sides are nearly equal, Betelgeuse being at the westernmost, and Procyon at the easternmost, angle.

351. Fig. 76 represents, in like manner, some of the equatorial constellations visible in the south on the 21st of May. Arcturus is now nearly on the meridian. East of it is the constellation Hercules, toward a point of which our Sun is travelling with his system of planets, satellites, and comets. Hercules may be recognized by the four-sided figure (nearly a square) formed by four of its brightest stars.

FIG. 76.—EQUATORIAL CONSTELLATIONS, VISIBLE IN THE SOUTH ON MAY 21, AT 10 P. M.

Apparent Movements of the Sun.

352. The effect of the Earth's daily movement upon the Sun is precisely similar to its effect on the stars; that is, the Sun appears to rise and set every day. But in

---

the heavens? 351. What does Fig. 76 represent? What star is now nearly on the horizon? What constellation is east of Arcturus? How may Hercules be recognized? 352. What is the effect of the Earth's daily movement on the Sun? How is the Sun's apparent motion affected by the Earth's yearly motion? 353.

SIDEREAL AND SOLAR DAY.

consequence of the Earth's yearly motion round it, it appears to revolve round the Earth more slowly than the stars; and it is to this that we owe the difference between star-time and sun-time, or between the sidereal and the solar day.

353. Difference between the Sidereal and the Solar Day.—How this difference arises is shown in Fig. 77, which represents the Sun, and the Earth in two positions in its orbit, separated by the time of a complete rotation. In the first position of the Earth are shown one observer, a, with the Sun on his meridian, and another, b, with a star on his; the two observers being on exactly opposite sides of the Earth, and on a line drawn through the centres of the Earth and Sun. In the second position, when the same star comes to b's meridian, a sees the Sun to the east of his, and he must be carried by the Earth's rotation to c before the Sun occupies the same apparent position in the heavens that it did before—or is again on his meridian. The solar day, therefore, will be longer than the sidereal day by the time it takes a to travel this distance. Of course, were the Earth at rest, this difference could not have arisen; the solar day is a result of the Earth's motion in its orbit, combined with its rotation.

FIG. 77.—EXPLANATION OF THE DIFFERENCE IN LENGTH BETWEEN THE SIDEREAL AND THE SOLAR DAY.

354. Moreover, the Earth's motion in its orbit is not uniform, as we shall hereafter see. Consequently, the

---

With Fig. 77, explain the difference between the sidereal and the solar day. By what is this difference caused? 354. Is the solar day always of the same length?

apparent motion of the Sun is not uniform, and solar days are not of the same length; for it is evident that if the Earth sometimes travels faster, and therefore farther, in the interval of one rotation than it does in another, the observer *a* has farther to travel before he gets to *c ;* and as the Earth's rotary motion is uniform, he requires more time. In a subsequent chapter it will be shown how this irregularity in the apparent motion of the Sun is obviated.

355. **Celestial Latitude and Longitude.**—The apparent yearly motion of the Sun is so important that astronomers map out the celestial sphere by a second method, in order to indicate his motion more easily; for as the plane of the celestial equator, like that of the terrestrial equator, does not coincide with the plane of the ecliptic, the Sun's distance from the celestial equator varies every minute.

To get over this difficulty, they make of the plane of the ecliptic a sort of second celestial equator. They apply the term *Celestial Latitude* to angular distances from it to the Poles of the Heavens, which are 90° from it north and south. They apply the term *Celestial Longitude* to the angular distance—reckoned on the plane of the ecliptic—from the position occupied by the Sun at the vernal equinox, reckoning from right to left up to 360°. This latitude and longitude may be either *heliocentric* or *geocentric,*—that is, reckoned from the centre of either the Sun or the Earth.

356. **The Zodiac and its Signs.**—The celestial equator in this second arrangement is represented by a circle called the Zodiac, which is divided, not only into degrees, etc., like all other circles, but also into Signs of 30° each. These, with their symbols, are as follows:—

Why not? 355. What have astronomers done, in order to indicate the Sun's motion more easily? What is meant by Celestial Latitude and Celestial Longitude? What is the meaning of the terms *heliocentric* and *geocentric,* as applied to celestial latitude and longitude? 356. In this arrangement, how is the celestial equator represented? How is the Zodiac divided? Name the spring signs; the summer signs; the autumn signs; the winter signs. With what must these

SIGNS OF THE ZODIAC.

Spring Signs.	Summer Signs.	Autumn Signs.	Winter Signs.
♈ Aries.	♋ Cancer.	♎ Libra.	♑ Capricorn.
♉ Taurus.	♌ Leo.	♏ Scorpio.	♒ Aquarius.
♊ Gemini.	♍ Virgo.	♐ Sagittarius.	♓ Pisces.

At the time this division was adopted, the Sun entered the constellation Aries at the vernal equinox, and traversed in succession the constellations bearing the above names; but at present, owing to the Precession of the Equinoxes, which will be explained hereafter, *the signs no longer correspond with the constellations*, and must not therefore be confounded with them.

357. **Relation between the Ecliptic and the Celestial Equator.**—These two methods of dividing the celestial sphere, and of determining the places of the heavenly bodies in it, refer, one to the plane of the terrestrial equator, and the other to the plane of the ecliptic. Now two things must be remembered:—(1) The angle formed by the celestial equator with the plane of the ecliptic is the same as that formed by the terrestrial equator,—that is, 23½° nearly. (2) The poles of the heavens are the same distance (23½°) from the celestial poles.

Moreover, if we regard the centre of the celestial sphere as lying at the centre of the Earth, it is clear that the two planes will intersect each other at that point; half of the ecliptic will be north of the celestial equator, and half below it; and there will be two points opposite to each other at which the ecliptic will cross the celestial equator.

358. **Apparent Path of the Sun.**—As the Sun keeps to the ecliptic, it must, at different parts of its path, cross the celestial equator, be north of it, cross it again, and be south of it; in other words, its latitude remaining the

---

signs not be confounded? Why were the names of the constellations given to them, if the signs and constellations do not agree? 357. What must be remembered with respect to the celestial equator and the poles of the heavens? If the centre of the Earth be taken as the centre of the celestial sphere, what will follow? 358. Describe the apparent path of the Sun. Give an account of its

same, its declination or distance from the celestial equator will change.

Hence, although the Sun rises and sets every day, its daily path is sometimes high, sometimes low. At the *vernal equinox*, when it occupies one of the points in which the ecliptic cuts the equator, it rises due east, and sets due west, like an equatorial star; then, as it gradually increases its north declination, its daily path approaches the zenith, and its rising and setting points advance northward, until it reaches that part of the zodiac at which the planes of the ecliptic and equator are most widely separated. Here it appears to stand still; we have the *summer solstice* (from the Latin *sol*, the sun, and *stare*, to stand), and its daily path is similar to that of a star of $23\frac{1}{2}°$ north declination. It then descends through the *autumnal equinox* to the *winter solstice*, when its apparent path is similar to that of a star of $23\frac{1}{2}°$ south declination, and its rising and setting points are low down toward the south.

359. **To determine the Time of Sunrise and Sunset with the Celestial Globe.**—The use of the celestial globe throws light on many points connected with the Sun's apparent motion. When we have rectified the globe, as directed in Art. 347, its top will represent the zenith,—a miniature terrestrial globe, with its axis parallel to that of the celestial one, being supposed to occupy the centre of the latter. By bringing different parts of the ecliptic to the brass meridian, the varying meridian height of the Sun, on which the seasons depend, is at once shown.

360. In addition to this, if we find from the almanac the position of the Sun in the ecliptic on any day, and bring it to the brass meridian, the globe shows the position of the Sun at noonday; the index-hand is, therefore, set to

---

movements after passing the vernal equinox. At the summer solstice. What does its apparent daily path resemble at the winter solstice? 359. How may the varying meridian height of the Sun be shown with the celestial globe? 360. How

12. If we then turn the globe westward till the Sun's place is brought close to the wooden horizon, we have sunset represented, and the index will indicate the time of sunset. If, on the other hand, we turn the globe eastward from the brass meridian till the Sun's place is brought close to the eastern edge of the wooden horizon, we have sunrise represented, and the index indicates the time of sunrise.

If the path of the Sun's place, when the globe is turned from the point occupied at sunrise to the point occupied at sunset, be carefully followed with reference to the horizon, the Diurnal Arc described by the Sun that day will be shown.

361. **To find the Length of Day and Night.**—At noon and midnight the Sun is mid-way between the eastern and western points of the horizon—part of his path being above the horizon, and part below it. The time, therefore, from noon to sunset is the same as from sunrise to noon. Similarly, the time from midnight to sunrise is equal to that from sunset to midnight.

As civil time divides the twenty-four hours into two portions, reckoned from midnight and noon, we have a convenient way of finding the length of the day and night from the times of sunrise and sunset. For instance, if the Sun rises at 7, the time from midnight to sunrise is seven hours; but this time is equal, as has been seen, to the time from sunset to midnight; therefore the night is fourteen hours long. Similarly, if the Sun sets at 8, the day is twice eight, or sixteen, hours long. Hence these rules:—

> Double the time of the Sun's setting is the length of the day.
>
> Double the time of the Sun's rising is the length of the night.

---

may the time of sunrise and sunset be determined? How may the Diurnal Arc described by the Sun be shown? 361. State two rules for finding the length of day and night from the time of sunrise and sunset. Give an example. Give the

Apparent Movements of the Moon.

362. The Moon, we know, makes the circuit of the Earth in $29\frac{1}{2}$ days; in one day, therefore, supposing her motion to be uniform, she will travel eastward over the face of the sky a space of about $12°$ ($360 \div 29\frac{1}{2} = 12\frac{1}{4}$ nearly). Accordingly, at a given hour, from night to night, her place will be changed about $12°$, and she rises and sets later in consequence.

Now, if the Moon's orbit were exactly in the plane of the ecliptic, we should not only have two eclipses every month (as heretofore stated), but she would appear always to follow the Sun's track. We have seen, however, that her orbit is inclined $5°$ to the plane of the ecliptic, and therefore to the Sun's apparent path. It follows, therefore, that when the Moon is approaching her descending node, her path dips down (and her north latitude decreases), and that when she is approaching her ascending node, her path dips up (and her southern latitude decreases).

The inclination of the Moon's orbit to the plane of the ecliptic being $5°$, the greatest possible difference between her meridian altitudes is twice the sum of $5°$ and $23\frac{1}{2}°$, or $57°$. That is to say, she may be $5°$ north of a part of the ecliptic which is $23\frac{1}{2}°$ north of the equator, or she may be $5°$ south of a part of the ecliptic which is $23\frac{1}{2}°$ south of the equator.

But let us suppose the Moon to move actually in the ecliptic. It is clear that the full Moon at midnight occupies exactly the opposite point in the ecliptic to that occupied by the Sun at noon-day. In winter, therefore, when the Sun is lowest, the Moon is highest; and so in winter we get more moonlight than in summer, not only because

---

reasoning on which these rules are based. 362. Why does the Moon rise and set later from night to night? If the Moon's orbit were exactly in the plane of the ecliptic, what would follow? How much is its orbit inclined to the plane of the ecliptic? What is the consequence? What is the greatest possible difference between the meridian altitudes of the Moon? At what season do we

the nights are longer, but because the Moon, like the Sun in summer, is best situated for lighting up the northern hemisphere.

363. **The Harvest Moon.**—Although, as we have seen, the Moon advances about 12° in her orbit every 24 hours, the interval between two successive moonrises varies considerably. If the Moon moved along the celestial equator, the interval would always be about the same, because the equator is always inclined the same to our horizon. But she moves nearly along the ecliptic, which is inclined $23\frac{1}{2}°$ to the equator; and because it is so inclined, she approaches the horizon at very different angles at different times, varying in lat. 40° between 21° and 79°.

In Art. 357 we saw that half of the ecliptic is to the north and half to the south of the equator; the ecliptic crosses the equator in the signs Aries and Libra. Now, when the Moon is farthest from these two points twice a month, her path is parallel to the equator, and the interval between two risings will be nearly the same for two or three days together; but mark what happens if she be near a node, *i. e.*, in Aries or Libra. In Aries the ecliptic crosses the equator to the north; in Libra, to the south. In Fig. 78 the line $H O$ represents the horizon, looking east; $E Q$ the equator, which in lat. 40° is inclined 50°

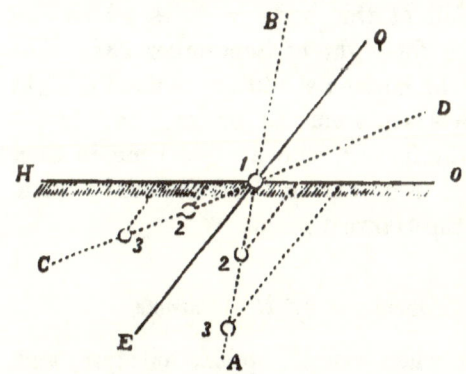

Fig. 78.—Explanation of the Harvest Moon.

---

get the most moonlight? Explain how this happens. 363. What is said of the interval between two successive risings of the Moon? Why does this interval vary? Under what circumstances is the time of the Moon's rising nearly the same for two or three successive days? Explain this with Fig. 78. How often

to the horizon. The dotted line AB represents the direction of the ecliptic when the sign Libra is on the horizon, and CD its direction when Aries is on the horizon.

Now, the Moon appears to rise because our horizon is carried down toward it. It follows that, when the Moon occupies the three positions shown on the line CD, she will rise nearly at the same time on successive evenings; though she has advanced each time 12° in her orbit, she has got very little farther below the horizon, as will be seen in the figure. On the other hand, on the line AB, her path being much more inclined to the horizon, each advance of 12° in her orbit carries her much farther below the horizon, and the difference between two successive risings will be greater in proportion. In lat. 40° this difference may be no more than 25 minutes, and on the other hand may amount to an hour and a quarter.

These successive risings of the Moon at nearly the same hour of course occur every month, as the Moon makes an entire circuit in a month and must pass the node in question in every circuit; but they are not noticed, except when the Moon is full at this node in Aries, which can happen only within a fortnight of September 23d. The full disk, seen above the horizon shedding its flood of light as soon as the Sun has set, seems to prolong the day,— most acceptably to the farmer, who at this time in England is busily engaged in gathering the fruits of the earth. Hence this is called the Harvest Moon.

Apparent Movements of the Planets.

364. The planets, when visible, appear as stars, and, like the stars, rise and set by virtue of the Earth's rotation. We need, therefore, to consider only their apparent

---

do these successive risings at nearly the same hour occur? When only are they noticed? What is the Moon called at this time? 364. How do the planets, when they are visible, appear? What apparent motions have they? Which of these

motions among the stars, caused by the Earth's revolution round the Sun, combined with their own actual movements.

365. **Distances of the Planets from the Earth.**—As the planets revolve round the Sun in orbits of very different size and at different rates of speed, their distances from each other and from the Earth are perpetually varying.

366. The Earth at one time has a given planet on the same side of the Sun as herself, and at another on the opposite side. The extreme distances, therefore, between the Earth and a superior planet will vary by the diameter of the Earth's orbit—that is, in round numbers, by 183,000,000 miles. In the case of an inferior planet, the extreme distances will differ by the diameter of the inferior planet's orbit. But this is not all; as the orbits are elliptical and the nearest approaches and greatest departures occur in different parts of them, the distance of any planet from the Earth even at these times will not always be the same.

367. The following table shows the average least and greatest distance of each planet from the Earth, leaving out of account the variation due to the ellipticity of the orbits. The first column presents the difference between the distances of each planet and the Earth from the Sun, and the second column gives their sum.

	Least Distance. Miles.	Greatest Distance. Miles.
Mercury,	56,038,000	126,823,000
Venus,	25,299,000	157,562,000
Mars,	47,882,000	230,742,000
Jupiter,	384,263,000	567,123,000
Saturn,	780,704,000	963,565,000
Uranus,	1,662,421,000	1,845,281,000
Neptune,	2,654,841,000	2,837,701,000

are we to consider? 365. What is said of the distances of the planets from each other and from the Earth? 366. What difference must there be between the greatest and least distance of a superior planet from the Earth? Of an inferior planet? What further affects the difference of distance? 367. What are shown in the table? How are the numbers in the first column obtained? How, those

368. To variations of distance are to be ascribed the striking changes of the planets in size and brilliancy at different times. The difference of size is greatest in the case of those planets whose orbits lie nearest that of the Earth, as shown in the table. Thus Venus, when nearest the Earth, appears six times larger than when it is farthest away, because it is really six times nearer to us; while the apparent size of Uranus and Neptune is hardly affected, as the diameter of the Earth's orbit is small compared with their distance from the Sun.

369. **Phases of the Planets.**—In the case of the planets which lie between us and the Sun, phases similar to those of the Moon are presented, because sometimes the planet is between us and the Sun, as is the case with the Moon when it is new; sometimes the Sun is between us and the planet, and consequently we see the illuminated hemisphere. At other times, as shown in Fig. 80, the Sun is to the right or left of the planet as seen from the Earth; and a part of both the bright and the dark hemisphere is presented to us. Among the superior planets, Mars is the only one that exhibits a marked phase, which resembles that of the gibbous Moon.

370. **Aspects of the Planets.**—By the Aspects of the planets are meant their positions in their orbits relatively to the Sun and the Earth. The aspects most frequently alluded to are Conjunction, Opposition, and Quadrature.

When an inferior planet is in a line between the Earth and Sun, it is said to be in *inferior conjunction* with the Sun; when it is in the same line, but beyond the Sun, it is said to be in *superior conjunction*.

When a superior planet is on the opposite side of the

In the second? 368. What changes in the planets are to be ascribed to variations in their distance from the Earth? In what planets are these changes greatest? Compare Venus with Neptune in this respect. 369. What planets exhibit phases, and why? What superior planet exhibits a marked phase, and what does it resemble? 370. What is meant by the Aspects of the planets? What aspects are most frequently alluded to? When is a planet said to be in *conjunction?*

ASPECTS OF THE PLANETS.

Sun,—that is, when the Sun is between us and it,—we say it is in *conjunction;* when in the same straight line, but with the Earth in the middle, we say it is in *opposition*, because it is then in the part of the heavens opposite to the Sun.

When a planet is 90° from the position it occupies in conjunction and opposition, it is said to be *in quadrature*.

Conjunction is denoted by the sign ☌ ; opposition, by the sign ☍ ; quadrature, by the sign ☐.

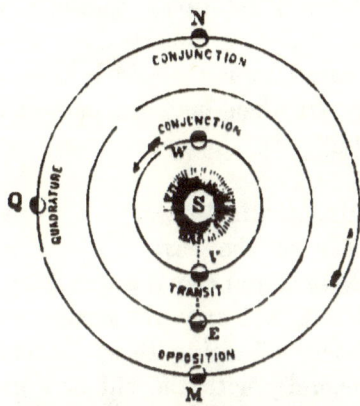

Fig. 79.—Conjunction—Opposition—Quadrature.

In Fig. 79, *S* represents the Sun, and *E* the Earth. *V* is Venus in inferior conjunction; *W* is the same planet in superior conjunction. Mars is shown in conjunction at *N*, in opposition at *M*, and in quadrature at *Q*.

371. **Transits.**—The passage of an inferior planet across the Sun's disk is called its Transit. In Fig. 79, Venus at *V* is making her transit.

A transit can take place only when a planet is in inferior conjunction. But, as the orbits of the planets do not lie in the plane of the ecliptic, there may be inferior conjunctions without any transit. Venus may be seen from the Earth in the same quarter as the Sun, and yet lie out of the plane which contains the centres of the Sun and the Earth.

372. **Elongations.**—If an observer could watch the motions of the planets from the Sun, he would see them all pursuing their courses, always in the same direction, with different velocities, but in the case of any particular

---

When, *in opposition?* When, *in quadrature?* By what signs are these aspects denoted? Illustrate these positions with Fig. 79. 371. What is a Transit? When alone can a transit take place? May there be inferior conjunctions without any transit? 372. What complicates the movements of the planets, as seen

one with an almost uniform rate of speed. Not only, however, is our Earth a moving observatory, the motion of which complicates the apparent movements of the planets in an extraordinary degree, but from its position in the system all the planets are not seen with equal ease.

In the first place, it is evident that only the superior planets are ever visible at midnight, as they alone can occupy the region of the heavens opposite to the Sun's place at that time, which is the region brought round to us at midnight by the Earth's rotation. Secondly, not only are the inferior planets always apparently near the Sun, but when they are nearest to us their dark sides are turned toward us, as they are then between us and the Sun, and the Sun is shining on the side turned away from us.

The greatest angular distance, in fact, of Mercury and Venus from the Sun, either to the east (left) or west (right) of it, called the Eastern and the Western Elongation, is, as heretofore stated, 29° and 47° respectively. Consequently, our only chance of seeing these planets is either in the day-time (generally with the aid of a good telescope), or just before sunrise at a western elongation, or after sunset at an eastern elongation.

373. **Stationary-points—Retrograde Motion.**—In Fig. 80 are shown the Earth in its orbit (P), and an inferior planet at its conjunctions and elongations. It is obvious that the rate and direction of the planet, as seen from the Earth, which for the sake of simplicity we will suppose to remain at rest, will both vary. At superior conjunction (SC) the planet will appear to move in the direction indicated by the outside arrow; when it arrives at its eastern

---

from the Earth? What planets alone are visible at midnight, and why? Where are the inferior planets always situated, and when are they invisible to us? What is meant by the Elongation of a planet? What is the greatest elongation of Mercury? Of Venus? What are our only times for seeing these planets? 373. What does Fig. 80 represent? With the aid of the figure, and supposing the Earth to be at rest, explain the apparent course of the inferior planet, its

STATIONARY-POINTS, RETROGRADATIONS. 207

FIG. 80.—ELONGATIONS, STATIONARY-POINTS, AND RETROGRADATIONS, OF THE PLANETS.

elongation (*EE*), it will appear to be *stationary*, because it is then for a short time travelling exactly toward the Earth. From this point, instead of journeying from right to left, as at superior conjunction, it will appear to us to travel from left to right, or *retrograde*, until it reaches the point of westerly elongation (*WE*), when for a short time it will travel exactly from the Earth, and again appear *stationary*, after which it recovers its direct motion.

The only difference made by the Earth's own movements in this case is, that, as its motion is in the same direction as that of the inferior planet, the intervals between two successive conjunctions or elongations will be longer than if the Earth were at rest.

374. The superior planets, as seen from the Earth, appear to reach stationary-points in the same manner, but for a different reason. At the moment a superior planet appears stationary, the Earth, as seen from that planet, has reached its point of eastern or western elongation. Let *P* in Fig. 80 represent a superior planet at rest, and let the inferior planet represented be the Earth. From the western elongation through superior conjunction, the motion of the planet referred to the stars beyond it will be *direct*—*i. e.* from *1 to *2, as shown by the outside

arrow. When the Earth is at its eastern elongation, as seen from the planet, the planet as seen from the Earth will appear at rest, as we are advancing for a short time straight to it. When this point is passed, the apparent motion of the planet will be reversed; it will appear to retrograde from *2 to *1, as shown by the inside arrow.

As in the former case, the only difference when we deal with the planet in motion, will be that the times in which these changes take place will vary with the actual motion of the planet; for instance, it will be much less in the case of Neptune than in that of Mars, as the former moves much more slowly.

375. **Synodic Period.**—In consequence of the Earth's motion, the period in which a planet regains the same position with regard to the Earth and Sun is different from the actual period of the planet's revolution round the Sun. The time in which a position, such as conjunction or opposition, is regained, is called a Synodic Period. The synodic periods of the different planets are as follows:—

	Mean Solar Days.		Mean Solar Days.
Mercury,	115.87	Saturn,	378.09
Venus,	583.92	Uranus,	369.66
Mars,	779.94	Neptune,	367.49
Jupiter,	398.87		

These synodic periods have been found by actual observation, and from them the times of the planets' revolution round the Sun have been obtained.

376. **Inclinations and Nodes of the Orbits.**—If the motions of the planets were confined to the plane of the ecliptic, they would, as seen from the Earth, resemble those of the Sun; but their orbits are all more or less inclined

---

a superior planet, supposed to be at rest. What difference will the motion of the planet make? 375. What is meant by a Synodic Period? Why does a planet's synodic period differ from its period of revolution round the Sun? State the synodic periods of the different planets. How have they been found? What have been obtained from them? 376. What causes the apparent motion of the

INCLINATION OF THE PLANETARY ORBITS.

to that plane (Art. 145). Here is a table of the present inclinations, and positions of the ascending nodes:—

Fig. 81.

	Inclination of Orbit.	Longitude of Ascending Node.
	° ′ ″	° ′
Mercury,	7 0 5	45 57
Venus,	3 23 29	74 51
Mars,	1 51 6	47 59
Jupiter,	1 18 52	98 25
Saturn,	2 29 36	111 56
Uranus,	0 46 28	72 59
Neptune,	1 46 59	130 6

377. The apparent distance of a planet from the plane of the ecliptic will be greater, as seen from the Earth, if the planet is nearer the Earth than the Sun at the time of observation. Hence, as the distance of the planet from the Earth must be taken into account, the distance above or below the plane of the ecliptic will not appear to vary so regularly when seen from the Earth, as it would do could we observe it from the Sun.

Of course, when the planet is at a node, it will always appear in the ecliptic.

378. **Path of Venus among the Stars.**—Fig. 81 represents the path of Venus, as seen from the Earth from April to October, 1868. A study of it should make what has been said about the

planets to differ from that of the Sun? 377. Why does not the distance of a planet from the plane of the ecliptic vary as regularly, when seen from the Earth, as it would do if seen from the Sun? When will a planet appear in the ecliptic?

apparent motions of the planets quite clear. From April to June the planet's north latitude is increasing, while the node and stationary-point—which in this case coincide, though they do not always do so—are reached about the 25th of June. The southern latitude rapidly increases, until, on the 9th of August, the other stationary-point is reached, after which the south latitude decreases.

379. **Effect of the Ellipticity and Inclination of the Orbit in the case of Mars.**—The apparent path of a planet, then, is affected by the motions of the Earth

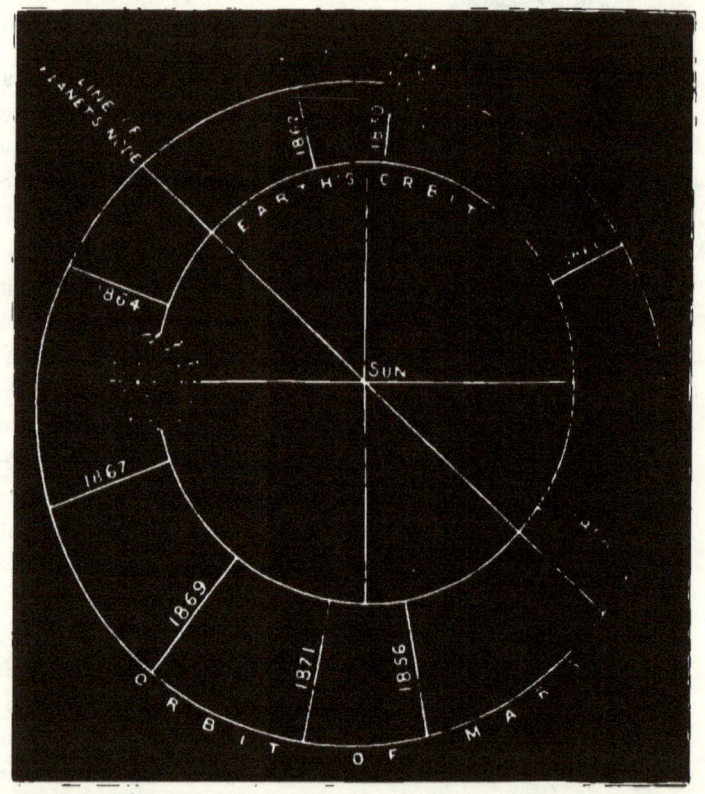

FIG. 82.—THE ORBITS OF MARS AND THE EARTH.

378. What does Fig. 81 represent? Describe the path of Venus, as thus ex

and the inclination of its own orbit. If we examine into the position of the orbit of Mars, for instance, more closely than we have hitherto done, we shall see how the ellipticity of the orbit and its inclination affect our observations of the physical features of the planet. Fig. 82 shows the exact positions in space of the orbits of the Earth and Mars, and the amount and direction of the inclination of their axes, and the line of the nodes of Mars; both planets are represented in the positions they occupy at the winter solstice of the northern hemisphere. The lines joining the two orbits indicate the positions occupied by both planets at successive oppositions of Mars, at which times, of course, Mars, the Earth, and the Sun, are in the same straight line (leaving the inclination of the orbit of Mars out of the question).

It is seen that at the oppositions of 1830 and 1860 the two planets were much nearer together than in 1867 or 1869.

380. Fig. 82 also enables us to understand that, in the case of an inferior planet, if we suppose the perihelion of the Earth to coincide in direction with (or, as astronomers put it, to be in the same heliocentric longitude as) the aphelion of the planet, the conjunctions which happen in this part of the orbits of both will bring the bodies nearer together than will the conjunctions which happen elsewhere. Similarly, if we suppose the aphelion of the Earth to coincide with the perihelion of a superior planet, as in the case of Mars, the opposition which happens in that part of the orbit will be the most favorable for observation. The Earth's orbit, however, is practically so nearly circular that the variation depends more upon the eccentricity of the orbits of the other planets than upon our own.

---

bibltcd. 379. By what is the apparent path of a planet affected? What does Fig. 82 represent? What is seen with respect to the nearness of the planets at different oppositions? 380. What may also be understood from Fig. 82? What

381. Fig. 82 also shows us that, when Mars is observed at the solstice indicated, we see the southern hemisphere of the planet better than the northern one; while at those oppositions which occur when the planet is at the opposite solstice, the northern hemisphere is chiefly visible. But we see more of the northern hemisphere in the latter case than we do of the southern one in the former, because in the latter case the planet is above the ecliptic, and we therefore see under it better; in the former it is below the ecliptic, and we see less of the southern hemisphere than we should do were the planet situated in the ecliptic.

FIG. 83.—DIFFERENT APPEARANCES OF SATURN'S RINGS.

382. **Saturn's Rings as seen at Different Times from the Earth.**—Fig. 83 shows the effect of inclination in the case of the rings of Saturn. The plane of the rings is inclined to the ecliptic, and the different positions of this

---

opposition will be the most favorable for observation? 381. When is the southern hemisphere of Mars best seen, and when the northern? Why at the latter time do we see the northern hemisphere more fully than we see the southern hemisphere at the former? 382. What is shown in Fig. 83? What different appear-

THE RINGS SOMETIMES INVISIBLE. 213

Fig. 84.—Appearance of Saturn when the Plane of its Rings passes through the Earth.

plane are always parallel. Twice in the planet's year the plane of the rings must pass through the Sun; and while the plane is sweeping across the Earth's orbit, the Earth, in consequence of its rapid motion, may pass two or three times through the plane of the ring.

Hence the ring-system about this time may be invisible, from three causes: (1) Its plane may pass through the Sun, and its extremely thin edge only will be lit up; (2) The plane may pass through the Earth; or (3) The Sun may be lighting up one surface, and the other may be presented to the Earth. These changes occur about every fifteen years, and in the mid-inter-

Fig. 85.—Saturn with the North Surface of its Rings presented to the Earth.

val the surface of the rings—sometimes the northern one, at others the southern—is presented to the Earth in the greatest angle.

In Fig. 63, page 148, Saturn was shown with the south surface of its ring-system presented to view. In Fig. 85, we have the aspect of the planet when the north surface of the ring is visible.

---

ances are presented by the rings? What three causes may render the ring-system invisible? How often do these changes occur? How is the ring-system presented to the Earth in the mid-interval? How is Saturn represented in Fig. 63? How, in Fig. 85?

CHAPTER XIII

THE MEASUREMENT OF TIME.

383. HAVING dealt with the apparent motions of the heavenly bodies, we now come to what those apparent motions accomplish for us,—namely, the division and exact measurement of time. For common purposes, time is measured by the Sun, as it is that body which gives us the primary division of time into day and night; but for astronomical purposes the stars are used, as the apparent motion of the Sun is subject to variation.

The correct measurement of time is not only one of the most important parts of practical Astronomy, but it is one of the most direct benefits conferred on mankind by the science; it enters, in fact, so much into every affair of life, that we are apt to forget that there was a period when that measurement was all but impossible.

384. **Clepsydræ and Sun-dials.**—Among the contrivances which were to the ancients what clocks and watches are to us, we may mention Clep'sydræ, or water-clocks, and Sun-dials. Of these, the former seem to have been the more ancient, and were used not only by the Greeks and Romans, but by other nations, the ancient Britons among them. In its simplest form it resembled the hour-glass, water being used instead of sand, and the flow of time being measured by the flow of the water.

After the time of Archimedes, clepsydræ of the most elaborate construction were common; but while they were in use, the days, both winter and summer, were divided into twelve hours from sunrise to sunset, and consequently the hours in winter were shorter than the hours in sum-

---

383. What does Chapter XIII. treat of? How is time measured for common purposes? How, for astronomical purposes? What is said of the importance of the correct measurement of time? 384 What instruments did the ancients use for measuring time? By whom were clepsydræ employed? Describe the

mer. The clepsydra, therefore, was almost useless except for measuring intervals of time, unless different ones were employed at different seasons of the year.

385. The sun-dial, also, is of great antiquity; it is referred to as in use among the Jews 742 B.C. This was a great improvement on the clepsydra; but at night and in cloudy weather it could not be used, and the rising, culmination, and setting of the various constellations, were the only means available for approximately telling the time during the night. Euripides, who lived 480–407 B.C., makes the Chorus in one of his tragedies ask the time in this form:—

"What is the star now passing?"

and the answer is,

"The Pleiades show themselves in the east;
The Eagle soars in the summit of heaven."

It is on record that as late as A.D. 1108 the sacristan of the Abbey of Cluny consulted the stars when he wished to know whether the time had arrived to summon the monks to their midnight prayers; in other cases, a monk remained awake, and to measure the lapse of time repeated certain psalms, experience having taught him in the day, by the aid of the sun-dial, how many psalms could be said in an hour.

To tell the passing hours, Alfred the Great (985 A.D.) used wax candles twelve inches in length. Marks on the surface at equal intervals denoted hours and their subdivisions, each inch of candle burnt showing that about twenty minutes had passed. To prevent currents of air from making his candles burn irregularly, he enclosed them in cases of thin transparent horn.

386. **Construction of the Sun-dial.**—To understand the construction of the sun-dial, let us imagine a transparent cylinder, having an opaque axis, both axis and cylinder being placed parallel to the axis of the Earth. If the

---

clepsydra in its simplest form. When did clepsydræ of elaborate construction become common? What difficulty interfered with their usefulness? 385. How early is the sun-dial known to have been in use? How did it compare with the clepsydra? How was the hour told at night? What question and answer occur in one of the tragedies of Euripides? How was the time for summoning the monks to their midnight prayers determined, as late as 1108 A.D.? To what

cylinder be exposed to the Sun, the shadow of the axis will be thrown on the side of the cylinder away from the Sun; and, as the Sun appears to travel round the Earth's axis in 24 hours, it will also appear to travel round the axis of the cylinder in the same time, and will cast the shadow of the axis on the side of the cylinder as long as it remains above the horizon.

All we have to do, therefore, is to trace on the side of the cylinder 24 lines 15° apart ($360 \div 24 = 15$), taking care to have one line

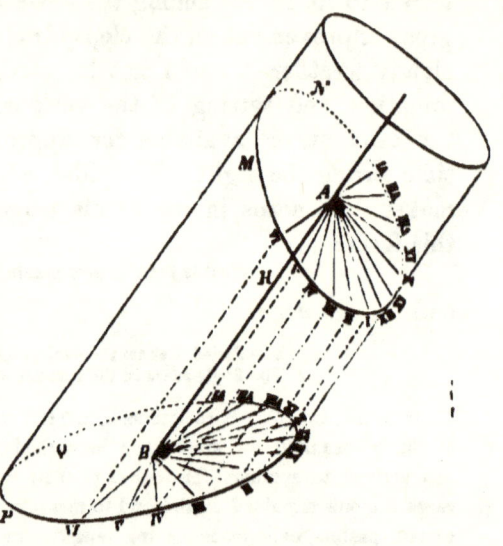

FIG. 86.—SUN-DIAL. *A B*, axis of cylinder; *M N, P Q*, two dials, at different angles to the plane of the horizon, showing how the imaginary cylinder determines the hour-lines.

due north of the axis. When the Sun is south, at noon, the shadow of the axis will be thrown on this line, which we mark XII. When the Sun has advanced 15° to the west, the shadow will be thrown on the next line to the east, which we mark I., and so on. The distance of the Sun above the equator will evidently make no difference in the lateral direction of the shadow.

387. In practice, however, we do not need the cylinder; all we want is a projection called a Style, parallel to the Earth's axis, and a Dial. The dial may be upright, horizontal, or inclined in any way so as to receive the shadow

---

device for measuring time did King Alfred resort? 386. Explain the construction of the sun-dial. 387. Do we really need a cylinder? What are needed

of the style; the lines on it indicating the hours will always be determined by imagining such a cylinder as is described above, cutting it parallel to the plane of the dial, and then joining the hour-lines on its surface with the style where it meets the dial.

388. **Clocks and Watches.**—The principle of both clocks and watches is that a number of wheels and pinions, working one in another, are forced to turn round, and are prevented from doing so too quickly. The force which gives the motion may be either a weight or a spring: the force which regulates the motion may proceed either from a pendulum, which at every swing locks the wheels, or from some equivalent arrangement.

389. Clocks appear to have been first used in Europe in the monasteries in the eleventh century; their invention is attributed to the Saracens. The first clock made in England, 1288 A. D., was considered so great a work that a high dignitary was appointed to take care of it, and paid for so doing from the public treasury.

Tycho Brahe used a clock, the motion of which was regulated by means of an alternating balance formed by suspending two weights on a horizontal bar, the movement being made faster or slower by altering the distances of the weights from the middle of the bar. But the clock, as an accurate measurer of time, dates from the middle of the seventeenth century, when the pendulum was introduced as a regulator by Galileo and Huyghens.

390. **The Mean Sun.**—In both clocks and watches we mark the flow of time by seconds, sixty of which make a minute, sixty minutes making an hour, and twenty-four hours a day. To the astronomer, however, the meaning

---

How may the dial be placed? 388. On what principle are both clocks and watches constructed? What is the force that imparts the motion? How is the motion regulated? 389. When were clocks first used in Europe? To whom is their invention attributed? When was the first clock made in England? How was it regarded? How did Tycho Brahe regulate his clock? When and by whom was the pendulum introduced as a regulator? 390. What does the word

of the word *day* is indefinite, unless it is specified whether a solar or sidereal day is intended. As commonly used, the term means neither; for when it was found that, in consequence of the irregularity of the Earth's motion in its orbit, the solar days differ in length, with the view of establishing a uniform measure of time for civil purposes, a civil day was made the average of all the solar days in the year. Our common day, therefore, is not measured by the *true Sun*, as a sun-dial measures it, but by what is called the *mean Sun*.

391. **Irregularities of the Sun's Apparent Daily Motion.**—Let us inquire into the motion of the imaginary mean Sun, by means of which the irregularities of the Sun's apparent daily motion are obviated.

In the first place, the real Sun's motion is in the ecliptic, and is variable. Secondly, the Sun crosses the equator twice a year at the equinoxes, at an angle of $23\frac{1}{2}°$, while midway between the equinoxes its path is almost parallel to the equator. Hence, its real motion being performed at different angles to the equator, its apparent motion will vary when referred to that line, being least rapid when the angle is greatest.

392. Let us first deal with the first cause—the inequality of the real Sun's motion. When the Earth is nearest the Sun, about Jan. 1st, the Sun appears to travel through $1° 1' 10''$ of the ecliptic in 24 hours; at aphelion, about July 1, the daily arc is reduced to $57' 12''$. The first thing to be done, therefore, is to give a constant motion to the mean Sun.

The real Sun passes through the entire circle, or $360°$, in 365d. 5h. 48m. 46s., or about 365.2422 days. Hence the mean distance traversed in one day will be as many de-

---

day, as commonly used, mean? By what is it measured? 391. Where does the real Sun move? Is its motion uniform or variable? What, besides, causes its motion to appear irregular? 392. What is the length of the daily arc traversed by the Sun at perihelion and at aphelion? Find the arc which the mean Sun

grees as 365.2422 is contained times in 360,—or a little more than 0.985 degrees. This distance, therefore, which equals 59' 8.33", is the arc which the mean Sun travels daily.

393. If the true Sun moved in the equator instead of in the ecliptic, a table showing how far the mean and true Sun are apart for every day in the year would at once enable us to determine mean time. But the true Sun moves along the ecliptic, while the mean Sun must be supposed to move along the equator, so that it may be carried evenly round by the Earth's rotation. This brings out the second cause of the inequality of the solar days.

At the solstices the true Sun moves almost parallel to the equator; at the equinoxes it crosses the equator at an angle of $23\frac{1}{2}°$, and, when its motion is referred to the equator, time is lost. This will be rendered evident if on a celestial globe we place wafers, equally distant from the first point of Aries, on both the equator and the ecliptic, and bring them to the brass meridian.

We have also the mean Sun, not supposed to move along the ecliptic at all, but along the equator, at the uniform rate of 59' 8.33" a day, and starting, so to speak, from the first point of Aries, where the ecliptic and equator intersect. Supposing the true Sun to move along the ecliptic at a uniform rate, its position referred to the equator would correspond with that of the mean Sun at the two solstices and the two equinoxes.

But the motion of the true Sun is not uniform; it moves fastest when the Earth is in perihelion, slowest when the Earth is in aphelion; and, if we take this fact into account, we find that the real Sun and the mean Sun coincide in

---

travels daily. 893. How does the true Sun move at the solstices? What does it do at the equinoxes? Does it appear to move more slowly or rapidly at the latter points? If the true Sun moved along the ecliptic at a uniform rate, at what points would it correspond with the mean Sun? Taking the irregular motion of the true Sun into account, at what times do the real Sun and the mean Sun coincide

position four times a year; namely, at April 15th, June 15th, August 31st, and December 24th.

394. **Equation of Time.**—At the following dates the difference between apparent and mean time is as specified below:—

	Minutes.		Minutes.
February 11th,	+ 14½	July 25th,	+ 6
May 14th,	− 4	November 1st,	− 16¼

This is what is called the Equation of Time, and is what we must add to, or subtract from, the time shown by a sun-dial, to make it correspond with that of a correct clock. The sign + before the equation of time denotes that it is to be added; the sign −, that it is to be subtracted.

When the Earth is in perihelion (or the Sun in perigee), the real Sun, moving at its fastest rate, gains on the mean Sun, and therefore takes longer than the mean Sun to come to the meridian; hence the dial is behind the clock, and we must add the equation of time to the apparent time to get the mean time. When the Earth is in aphelion (or the Sun in apogee), the reverse holds good. In November, as shown by the above table, the true Sun sets 16m. earlier than it would do if it occupied the position of the mean Sun, by which our clocks are regulated. In February it sets 15m. later: hence at the beginning of the year the day lengthens more rapidly than it would otherwise do. We cannot obtain mean time at once from observation; but, from an observation of the true Sun, by adding or subtracting the equation of time, as the case may be, it can be readily deduced. Mean time is now universally used in all civilized countries.

395. **Commencement of the Different Days.**—We must next consider when the different days begin. We have,

---

in position? 894. What is meant by the Equation of Time? What do the signs + and − mean, when prefixed to the equation of time? When is the equation of time to be added, and when subtracted? What is the equation of time for Feb. 11th? For May 14th? For July 25th? For Nov. 1st? How is

I. The **Apparent Solar Day**, reckoned from the instant the true Sun crosses the meridian, till it crosses it again.

II. The **Mean Solar Day**, reckoned from the instant the mean Sun crosses the meridian, in the same manner. Both these days are used by astronomers.

III. The **Civil Day**, commencing at midnight, and reckoned through 12 mean hours only to noon, and thence through another 12 hours to the next midnight. The civil reckoning is therefore always 12 hours in advance of the astronomical reckoning; hence this rule for determining the latter from the former:—For P. M. civil times, make no change; but for A. M. diminish the day of the month by 1 and add twelve to the hours. Thus: Jan. 2d, 7h. 49m. P. M. civil time, is Jan. 2d, 7h. 49m. astronomical time; but Jan. 2d, 7h. 49m. A.M. civil time is Jan. 1st, 19h. 49m. astronomical time.

396. **Length of the Different Days.**—Expressed in mean time, the length of the day is as follows:—

Apparent solar day, . . . variable.
Mean solar day, ... 24h. 0m. 0s.
Sidereal day, ... 23 56 4.09
Mean lunar day, . . . 24 54 0

397. **Sidereal Time** is reckoned from *the first point of Aries*. When the mean Sun occupies this point, which it does at the vernal equinox, the mean-time clock and the sidereal clock will agree. But this happens at no other time, as the sidereal day is only 23h. 56m. 4s. (mean time) long; so that the sidereal clock gains about four minutes a day, or one day a year, as compared with mean time. Of

---

mean time obtained? 395. When does the apparent solar day begin? The mean solar day? When does the civil day begin, and how is it reckoned? How does the civil reckoning compare with the astronomical reckoning. Give the rule for changing civil time to astronomical time. Give an example. 896. What is the length of the apparent solar day? Of the mean solar day? Of the sidereal day, in mean time? Of the mean lunar day? 397. From what is sidereal time reckoned? When will the mean-time clock and the sidereal clock agree? When

course the coincidence is established again at the next vernal equinox.

A sidereal clock represents the rotation of the Earth on its axis, as referred to the stars, its hour-hand performing a complete revolution through the 24 sidereal hours between the departure of any meridian from a star and its next return to it. At the moment that the vernal equinox, or a star whose right ascension is 0h. 0m. 0s. is on the meridian of Greenwich, the sidereal clock ought to show 0h. 0m. 0s.; and at the succeeding return of the star, or the equinox, to the same meridian, the clock ought to indicate the same time.

398. **The Week.**—Although the week, unlike the day, month, and year, is not connected with the movements of any heavenly body, the names of the seven days of which it is composed were derived by the Egyptians from the seven celestial bodies then known. The Romans, in their names for the days, observed the same order, distinguishing them as follows:—

Dies Saturni,	Saturn's day,	Saturday.
Dies Solis,	Sun's day,	Sunday.
Dies Lunæ,	Moon's day,	Monday.
Dies Martis,	Mars' day,	Tuesday.
Dies Mercurii,	Mercury's day,	Wednesday.
Dies Jovis,	Jupiter's day,	Thursday.
Dies Veneris,	Venus's day,	Friday.

We see at once the origin of our English names for the first three days; the remaining four are named from Tiw, Woden, Thor, and Frigga, northern deities equivalent to Mars, Mercury, Jupiter, and Venus, in the classical mythology.

---

will they next agree? Why do they not agree meanwhile? What does a sidereal clock represent? 398. From what did the Egyptians name the days of the week? Who observed the same order in their names for the days? Give the Latin names for the days of the week. Whence are their English names derived?

THE MONTH.

399. **The Month.**—We next come to the month, a period regulated entirely by the Moon's motion round the Earth.

The lunar month is the same as the *lunation* or *synodic month*, and is the time which elapses between two consecutive new or full Moons, or in which the Moon returns to the same position relatively to the Earth and Sun.

The tropical month is the revolution of the Moon with respect to the movable equinox.

The sidereal month is the interval between two successive conjunctions of the Moon with the same fixed star.

The anomalistic month is the time in which the Moon returns to the same point (for example, the perigee or apogee) of her movable elliptic orbit.

The nodical month is the time in which the Moon accomplishes a revolution with respect to her nodes, the line of which is also movable.

The calendar month is the month recognized in the almanacs, and consists of different numbers of days, such as January, February, etc.

400. **Length of the Lunar and other Months.**—The length of these different months is as follows:—

	Mean Time.			
	d.	h.	m.	s.
Lunar, or Synodic month,	29	12	44	2.84
Tropical month,	27	7	43	4.71
Sidereal month,	27	7	43	11.54
Anomalistic month,	27	13	18	37.40
Nodical month,	27	5	5	35.60

401. **The Year.**—The *year* is the time of the Earth's revolution round the Sun, as the *day* is the period of its rotation on its axis. There are various sorts of years, as there are different kinds of days. Thus, we may take the

---

399. By what is the month regulated? What is the Lunar Month? The Tropical Month? The Sidereal Month? The Anomalistic Month? The Nodical Month? The Calendar Month? 400. Which is the longest of these different kinds of months? Which is the shortest? 401. What is the Year? What is the Sidereal Year? The Solar, or Tropical, Year? The Anomalistic Year? Which is the

time that elapses between two successive conjunctions of the Sun, as seen from the Earth, with a fixed star. This is called the Sidereal Year.

Or we may take the period that elapses between two successive passages through the vernal equinox. This is called the Solar, or Tropical Year, and it is shorter than the sidereal year, in consequence of the precession of the equinoxes. The vernal equinox in its recession meets the Sun, which therefore passes through it sooner than it would otherwise do.

Again, we may take the time that elapses between two successive passages of the Earth through perihelion or aphelion. As these points have a forward motion in the heavens, the Anomalistic Year, as this period is called, is longer than the sidereal year.

402. **Length of the Sidereal and other Years.**—The exact length of these years is as follows:—

	Mean Time.			
	d.	h.	m.	s.
Mean sidereal year,	365	6	9	9.6
Mean solar or tropical year,	365	5	48	46.05444
Mean anomalistic year,	365	6	13	49.3

403. **The Calendar.**—It is seen from this table that the solar year does not contain an exact number of solar days, but nearly a quarter of a day over. It is said that the inhabitants of ancient Thebes were the first to discover this. The calendar had got in such a state of confusion in the time of Julius Cæsar, that he called in the aid of the Egyptian astronomer, Sosigenes, to reform it. The latter recommended that one day every four years should be added, by reckoning the sixth day before the kalends of March (Feb. 24th) twice; hence the term Bissextile (from the Latin *bis*, twice, and *sextus*, sixth).

---

longer, the solar or the sidereal year? The anomalistic or the sidereal year? 402. What is the exact length of the mean solar year? 403. What caused the calendar to get in confusion in old times? Who attempted to reform it? Whom

Now, this arrangement was a great improvement; but too much was added, and the matter was again looked into in the sixteenth century, by which time the over-correction had amounted to more than ten days, the vernal equinox falling on March 11th, instead of March 21st. Pope Gregory XIII., therefore, undertook to continue the good work begun by Julius Cæsar, and made the following rule for the future:—Every year divisible by 4 (except the secular years, 1800, 1900, etc.) to be a bissextile, or leap-year, containing 366 days; every year not so divisible to consist of only 365 days; every secular year divisible by 400 to be a leap-year; every secular year not so divisible to consist of 365 days. According to this arrangement, the error amounts to only 1 day in 3,866 years.

404. **Old and New Style.**—The Julian Calendar (Julius Cæsar's) was introduced 46 B. C.; the Gregorian (Pope Gregory's), in 1582 A. D. The latter was not adopted in England till 1752, when the correction was made by dropping eleven days in September, the day following the 2d of that month being called the 14th. This was known as the New Style (N. S.), in contradistinction to the Old Style (O. S.). In Russia the old style is still retained, although it is customary to give both dates, thus: 1870 $\frac{\text{Aug. 23}}{\text{Sept. 4}}$.

405. It is all-important that the calendar be exactly adjusted to the length of the solar year; otherwise the seasons would not commence on the same day of the same month as they do now, but would in the course of time make the circuit of all the days in the year. January, or any other month, would fall successively in spring, summer, autumn, and winter.

---

did Cæsar call to his aid? What improvement was made by Sosigenes? What difficulty still remained? By whom was this obviated? What change was made by the Gregorian Calendar? According to this calendar, what is the amount of error? 404. When was the Julian Calendar introduced? When, the Gregorian? When was the Gregorian Calendar adopted in England? How was the correction made? How were the two modes of reckoning distinguished? Where is the Old Style still retained? 405. Why is it important that the calendar should be

406. **Change in the Length of the Solar Year.**—At present, owing to a change of form in the Earth's orbit, the solar year is diminishing at the rate of $\frac{1}{10}$ of a second in a century. It is shorter now than it was in the time of Hipparchus by about 12 seconds.

407. **Change of Aphelion and Perihelion.**—If the solar and the anomalistic year were of equal length, it would follow that, as the seasons are regulated by the former, they would always occur in the same part of the Earth's orbit. As it is, however, the line joining the aphelion and perihelion points, termed the Line of Apsides, slowly changes its direction, at such a rate that in a period of 21,000 years it makes a complete revolution. At present, as already stated, we are nearest to the Sun about Jan. 1st; in A. D. 6485, the perihelion will correspond with the vernal equinox.

CHAPTER XIV

ASTRONOMICAL INSTRUMENTS.

Light.

408. **What Light is.**—Modern science teaches us that Light consists of undulations or waves of a medium called *ether*, which pervades all space. These undulations are to the eye what sound-waves are to the ear, and they are set in motion by bodies at a high temperature—the Sun, for instance—much in the same manner as the air is put in motion by our voice, or the surface of water by throwing in a stone. But though a wave-motion results from all

---

exactly adjusted to the length of the solar year? 406. At what rate is the solar year constantly changing, and why 407. What is meant by the Line of Apsides? What change is this line undergoing? When are we at present nearest the Sun? When will the perihelion correspond with the vernal equinox?

408. Of what does Light consist? To what are waves of light analogous?

these causes, the way in which the wave travels varies in each case.

409. **Velocity of Light.**—Though light moves so quickly that to us its passage seems instantaneous, it requires time to travel from an illuminating to an illuminated body. Its velocity was determined by Roemer, a Danish astronomer, from observations on the moons of Jupiter. He found that the eclipses of these moons (which he had calculated beforehand) happened 16m. 26s. later when Jupiter was in conjunction with the Sun than when he was in opposition. Knowing that Jupiter is farther from us in the former case than in the latter, by exactly the diameter of the Earth's orbit, he soon convinced himself that the difference of time was due to the fact that the light had so much farther to travel. Now, the additional distance, *i. e.*, the diameter of the Earth's orbit, being 183,000,000 miles, it follows that light travels about 185,000 miles a second. This fact has been abundantly proved since Roemer's time, and what astronomers call the *aberration of light* is one of the proofs.

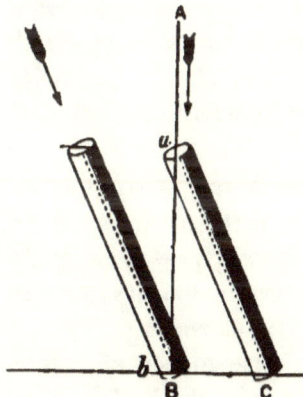

Fig. 87.—Illustrating the Aberration of Light.

410. **Aberration of Light.**—We may get an idea of the aberration of light by observing the way in which, when caught in a shower, we hold the umbrella inclined in the direction in which we are hastening, instead of overhead, as we should do were we standing still. Let us make this a

---

409. By whom was the velocity of light determined? How was it determined? What is the velocity of light? What is one proof of the velocity thus established?
410. How may we get an idea of the aberration of light? Illustrate it with Fig. 87. To see a star, what must we do with our telescopes? On what does the

little clearer. Suppose we wish to let a drop of water fall through a tube (see Fig. 87) without wetting the sides. If the tube is at rest, there is no difficulty—it has only to be held upright in the direction AB; but if we must move the tube, the matter is not so easy. The diagram shows that the tube must be *inclined*, or else the drop in the centre of the tube at a will no longer be in the centre at b; and the faster the tube is moved, the more it must be inclined.

Now, we may liken the drop to rays of light, and the tube to the telescope, and we find that to see a star we must incline our telescopes in like manner. By virtue of this, each star really seems to describe a small circle in the heavens, representing on a small scale the Earth's orbit; the extent of this apparent circular motion depending upon the relative velocity of light and of the Earth in its orbit, as in Fig. 87 the slope of the tube depends on the relative rapidity of the motion of the tube and the drop. From the actual dimensions of the circle, we learn that light travels about 10,000 times faster than the Earth does—that is, about 185,000 miles a second. This velocity has been experimentally proved by Foucault, by means of a turning mirror.

411. **Reflection and Refraction.**—A ray of light is *reflected* by opaque bodies which lie in its path, and is *refracted*, or bent out of its course, when it passes obliquely from a transparent medium of a certain density, such as air, into another of a different density, as water.

412. **Effect of Refraction.**—In consequence of refraction, the stars appear to be higher above the horizon than they really are. In Fig. 88, AB represents a pencil of light coming from a star. In its passage through our at-

---

extent of the apparent circular motion of the star depend? From the actual dimensions of the circle, how fast is light found to travel? How has Foucault experimentally proved this velocity? 411. By what is a ray of light reflected? Under what circumstances is it refracted? 412. How do the stars appear, in

REFRACTION OF LIGHT.

mosphere, since each layer gets denser as the surface of the Earth is approached, the ray is gradually refracted until it reaches the surface at C; from which point the star seems to lie in the direction CB.

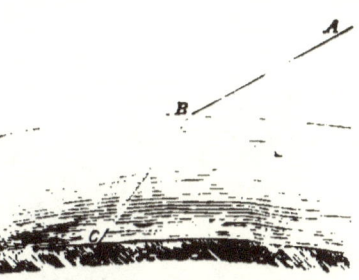

FIG. 88.—ILLUSTRATING REFRACTION.

413. The refraction of light can be best studied by means of a piece of glass with three rectangular faces, called a Prism. If we take a prism into a dark room, admit a beam of sunlight through a hole in the shutter, and let it fall obliquely on one of the surfaces of the prism, we shall see at once that the direction of the ray is changed. In other words, the angle at which the light falls on the first surface of the prism is different from the angle at which it leaves the second surface.

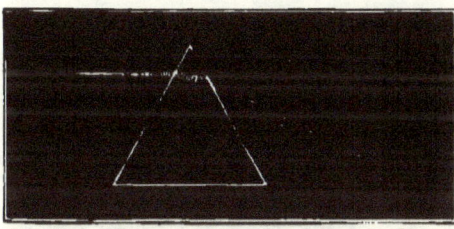

FIG. 89.—A PRISM, REFRACTING A RAY OF LIGHT.

414. **Dispersion of Light.**—If we receive a beam thus refracted by its passage through a prism, on a piece of smooth white paper, we shall have, instead of a spot of white light of the size of the hole that admitted the beam, a lengthened figure made up of seven different colors (as shown in Fig. 90), called the **Spectrum.**

By passing through the prism, the beam has been decomposed into colored rays, occupying different places on account of their different degrees of refrangibility, red

---

consequence of refraction? Illustrate this with Fig. 88. 413. How is refraction best studied? How can refraction be shown with a prism? 414. What is meant by the Spectrum? Mention the colors of the spectrum in order. Why do the colored rays occupy different places? Which colored ray is refracted the most?

230   ASTRONOMICAL INSTRUMENTS.

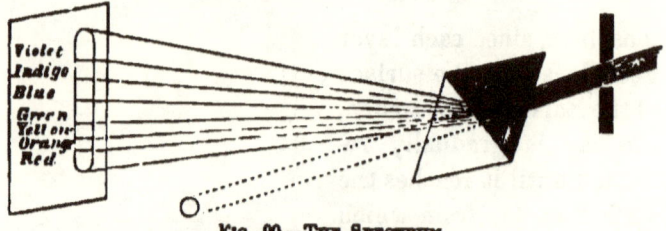

FIG. 90.—THE SPECTRUM.

being caused to deviate the least from the course of the original beam, and violet the most. This separation of light into the different colors of the spectrum is called Dispersion.

By passing the decomposed beam through a second prism placed in contact with the first, as in Fig. 90, the colored rays may be brought together again into a beam of white light.

415. If we pass light through prisms of different materials, we shall find that, although the colors always maintain the same order, they will vary in length. Thus, if we employ a hollow prism filled with oil of cassia, we shall obtain a spectrum two or three times longer than if we use one made of common glass. This fact is expressed by saying that different media have different dispersive powers.

Lenses.

416. A Lens is a transparent body (commonly of glass) which has two polished surfaces, either both curved or one curved and the other plane. The general effect of lenses is to refract rays of light, and magnify or diminish objects seen through them.

417. There are four kinds of lenses with which we have mainly to do; viz.,

---

Which, the least? How may the colored rays be brought together again into a beam of white light? What is the separation of light into the colors of the spectrum called? 415. If we pass light through prisms of different materials, what shall we find? Give an illustration. How is this fact expressed? 416. What is a Lens? What is the general effect of lenses? 417. With how many

LENSES.   231

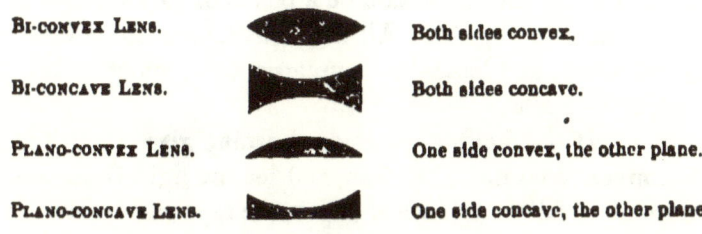

Bi-convex Lens.   Both sides convex.
Bi-concave Lens.   Both sides concave.
Plano-convex Lens.   One side convex, the other plane.
Plano-concave Lens. One side concave, the other plane.

Fig. 91. — Different Kinds of Lenses.

418. **Refraction by Convex Lenses.**—A prism refracts a ray of light as shown in Fig. 89; hence, two prisms arranged as in Fig. 92 would cause two parallel beams coming from different points at *a* and *b*, to *converge* at one point *c*.

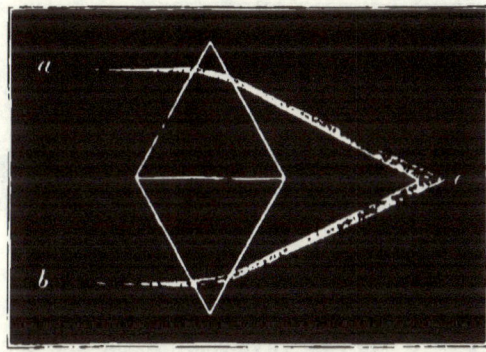

Fig. 92.—Action of two Prisms placed base to base.

We may look upon a bi-convex lens as composed of an infinite number of prisms; it will have a similar effect to that shown in Fig. 92. A section

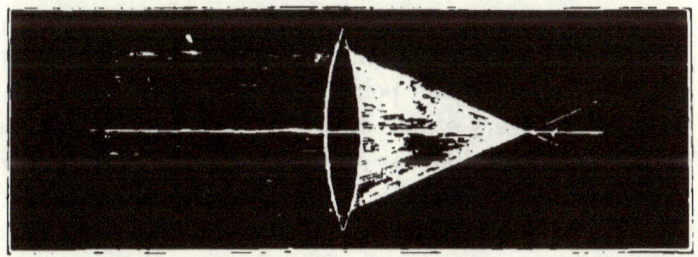

Fig. 93. —Bi-convex Lens, causing Parallel Rays to converge to a Focus.

kinds of lenses have we mainly to do? Name and describe them. 418. How are two prisms arranged in Fig. 92? What is their effect on two parallel beams? How may we regard a bi-convex lens? What will be the action of such a lens on

of such a lens and its action on a pencil of parallel rays are represented in Fig. 93. All the light falling on its surface is refracted, and made to converge to *c*, which point is called the **Focus**.

419. If we hold a common burning-glass (which is a bi-convex lens) up to the Sun, and let the light that passes through it fall on a piece of paper, the rays will be brought to a focus; and if the paper is held at a certain distance from the lens, a hole will be burned through it. This distance marks the Focal Distance of the lens.

420. If we place an arrow *a b* in front of the bi-convex lens *m n*, we shall have an image of the arrow behind the

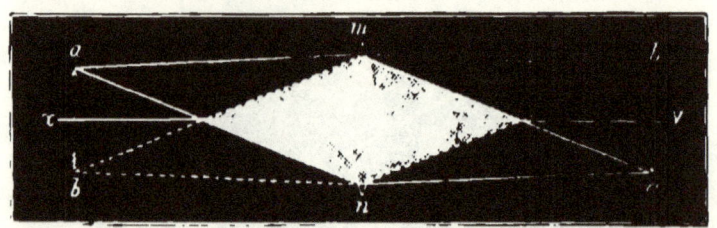

Fig. 91.— Bi-convex Lens, throwing an Inverted Image.

lens at *b a*, every point of the arrow sending a ray to every point in the surface of the lens. Each point of the arrow, in fact, is the apex of a cone of rays resting on the lens, and a similar cone of rays, after refraction, paints every point of the image. At *a*, for instance, in front of the lens, we have the apex of a cone of rays, *n a m*; which rays, being refracted, form another cone of rays, *n a m*, behind the lens, painting the point *a* in the image. So with *b*, and so with every other point. We see that the effect of a bi-convex lens, like the one in the figure, is to *form an inverted image*. The line *x y* is called the *axis* of the lens.

421. Such is the action of a bi-convex lens; and such a

---

a pencil of parallel rays? What is meant by the Focus? 419. What kind of a lens is a burning-glass? What is the effect of a burning-glass? What is meant by the focal distance of the lens? 420. Explain the action of a bi-convex lens in forming an image. What kind of an image does it form? 421. Explain the

lens we have in our eye. Behind it, where the image is cast, as in Fig. 94, we have a membrane which receives the image as the photographer's ground glass or prepared paper does; and when the image falls on this membrane, which is called the *ret'ina*, the optic nerves telegraph, as it were, an account of the impression to the brain, and *we see*.

422. In order that we may see, it is essential that the rays should enter the eye parallel or nearly so. Hence the use of the common magnifying-glass. We bring the glass close to the eye, and place the object to be magnified in its focus,—that is, at *c* in Fig. 93; the rays which diverge from the object are rendered parallel by the lens, and we are enabled to see the object, which appears large because it is brought so close to us.

423. **Refraction by Concave Lenses.**—If, instead of arranging the prisms as shown in Fig. 92, with their bases together, we place them point to point, it is evident that the rays falling upon them will no longer converge; they will in fact separate, or *diverge*. We may suppose a lens formed of an infinite number of prisms, joined together in this way. Such a lens is called a bi-concave lens. Its shape and action on parallel rays are shown in Fig. 95.

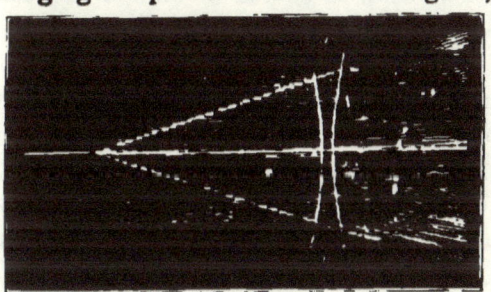

FIG. 95.—BI-CONCAVE LENS, CAUSING PARALLEL RAYS TO DIVERGE.

424. **Achromatic Lenses.**—A lens being equivalent, as we have seen, to a combination of prisms, we would natu-

---

principle on which we see. 422. In order that we may see, what is essential? Explain the principle of the common magnifying-glass. 423. Explain the action of a bi-concave lens. 424. What kind of an image would we naturally expect a

rally expect it to throw a colored image. This it does; and unless we could get rid of the colors, it would be impossible to make a large telescope worth using. By combining, however, two lenses of different shapes, and made of different kinds of glass, we cause the color to disappear, thus forming what is called an Achromatic Lens (from the Greek *a, without,* and χρῶμα, *color.*

425. We are able to get rid of color in the image in consequence of the varying dispersive powers (Art. 415) of different bodies. If we take two exactly similar prisms of the same material, and place one against the other as shown in Fig. 90, a beam of light passing through both will be unaffected; one prism will exactly undo the work done by the other, and the ray will be neither refracted nor dispersed. But if we take away the second prism, and replace it with one made of a substance having a higher dispersive power, we shall of course be able to counteract the dispersive effect of the first prism with a smaller thickness of the second.

But this smaller thickness will not counteract all the refractive effect of the first prism. The beam will therefore leave the second prism colorless, but refracted; and this is exactly what is wanted. The Chromatic Aberration, as it is called, is corrected, but the compound prism can still refract.

426. An achromatic lens is made in the same way as an achromatic prism. The dispersive powers of flint and crown glass are as .052 to .033. The front or convex lens is made of crown-glass. Its chromatic aberration is corrected by a bi-concave lens of flint-glass placed behind it. The second lens is not so concave as the first is convex; hence the refractive effect of the latter is not wholly

---

lens to form, and why? What would be the consequence, if we could not get rid of the color? How do we get rid of it? What is such a combination called? 425. How is the chromatic aberration, as it is called, corrected in the case of a prism? 426. How is an achromatic lens made? When is the spherical aberration

nullified. But as the second lens equals the first in dispersive power, although it cannot restore the ray to its original direction, it makes it colorless, or nearly so. If such an achromatic lens be truly made, and its curves properly regulated, it is said to have its *spherical* aberration corrected as well as its *chromatic* aberration, and the image of a star will form a nearly colorless point at its focus.

The Telescope.

427. **History.**—The Telescope, to which Astronomy is mainly indebted for the important advances it has made during the last two centuries, is an instrument for viewing distant objects. It appears to have been invented by Metius, a native of Holland, in 1608. Galileo, hearing of the invention, constructed an instrument for himself, and was the first to turn the telescope to practical account. Since his time, many improvements have been made, greatly increasing the efficiency of the instrument.

428. **Construction.**—The telescope is a combination of lenses. The principle involved in its construction is simply an extension of that exhibited in the structure of the eye. In the eye, nearly parallel rays fall on a lens, and this lens throws an image. In the telescope, nearly parallel rays fall on a lens, this lens throws an image, and then *another lens enables the eye to form an image of the image* by rendering the rays again parallel. These parallel rays enter the eye just as the rays do in ordinary vision.

In Fig. 96, for instance, let A represent the front lens, called the *object-glass*, because it is nearest to the object viewed; let C represent the other, called the *eye-piece*, because it is nearest the eye; and let B represent the image of a distant arrow, the rays from which are seen

---

said to be corrected, as well as the chromatic aberration? 427. To what is Astronomy mainly indebted for its recent advances? By whom was the telescope invented? By whom was it first used? 428. What is the principle involved

falling on the object-glass from the left. These rays are refracted, and we get an inverted image at the focus of the object-glass, which is also the focus of the eye-piece. The rays leave the eye-piece adapted for vision as they are when they fall on the object-glass; the eye can therefore use them as well as if no telescope had been there.

429. Illuminating Power.—The efficiency of the telescope depends on two things, its Illuminating and its Magnifying Power.

First, as to its Illuminating Power. The object-glass, being larger than the pupil of our eye, receives more rays than the pupil. If its surface be a thousand times greater than that of the pupil, for instance, it receives a thousand times more light; and consequently the image of a star formed at its focus is nearly a thousand times brighter than that thrown by the lens of our eye on the retina. We say *nearly* a thousand times, because some light is lost by reflection from the object-glass and during the passage through it. If we have two object-glasses of the same size, one highly polished and the other less so, the illuminating power of the former will be the greater.

Fig. 96.—Construction of the Astronomical Telescope.

---

in the construction of the telescope? Explain this further with Fig. 96. 429. On what does the efficiency of the telescope depend? How does the telescope get its illuminating power? How is some of the light that falls on the object-glass

THE TELESCOPE. 237

430. **Magnifying Power.**—The Magnifying Power depends upon two things. First, it depends upon the focal length of the object-glass; because, if we suppose the focus to lie in the circumference of a circle having its centre in the centre of the lens, the image will always bear the same proportion to the circle. Suppose it covers 1°; it is evident that it will be larger in a circle whose radius is 12 feet than in one whose radius is 12 inches—that is, in the case of a lens whose focal length is 12 feet, than in one whose focal length is 12 inches.

Next, the magnifying power of the eye-piece is to be taken into account. This varies according to the eye-piece used, the ratio of the focal length of the object-glass to that of the eye-piece giving its exact amount. Thus, if the focal length of the object-glass is 100 inches, and that of the eye-piece one inch, the telescope will magnify 100 times. But, unless the illuminating power is good and a perfect image is formed, a high magnifying power is useless. If the object-glass does not perform its part properly, the image will be blurred even when slightly magnified.

431. **Eye-pieces.**—The eye-pieces used with the astronomical telescope vary in form. The telescope made by Galileo, similar in construction to the modern opera-glass, was furnished with a bi-concave eye-piece. This eye-piece is introduced between the object-glass and the focus, at a point where its divergent action corrects the convergent effect of the object-glass, and thus makes the rays parallel. A convex eye-piece for the same reason is placed beyond the focus, as shown in Fig. 96.

Such eye-pieces, however, color the light coming from the image, in the same way as the object-glass would color

---

lost? 430. On what two things does the magnifying power of the telescope depend? Show how the focal length of the object-glass has to do with the magnifying power. What exactly shows the amount of magnifying power? With any magnifying power, what is essential? 431. Describe the eye-piece and its position in Galileo's telescope. What difficulty did the use of such eye-pieces involve? How did Huyghens remedy this difficulty? How are the plano-convex

the light which forms the image, if its chromatic aberration were not corrected.

It was discovered by Huyghens that this defect might be obviated by using two plano-convex lenses, the flat sides toward the eye,—the larger, called the field-lens, nearer the image, and the smaller, called the eye-lens, nearer the eye. This is the construction now generally used except in micrometers, in which the flat sides of the lenses are turned away from the eye.

432. The telescope-tube keeps the object-glass and the eye-piece in their proper positions; and the eye-piece is furnished with a draw-tube, which allows its distance from the object-glass to be varied.

433. **The Largest Refractor.**—The largest refracting telescope in the world is one recently constructed in England, having an object-glass 25 inches in diameter. The pupil of the eye is $\frac{1}{5}$ of an inch in diameter; this object-glass, therefore, will grasp over 15,000 (25 ÷ $\frac{1}{5}$ = 125; 125^2 = 15,625) times more light than the eye can. If used when the air is pure, it bears a power of 3,000 on the Moon; in other words, the Moon seen through it appears as it would were it 3,000 times nearer to us, or at a distance of 80 miles, instead of 240,000.

434. **Reflecting Telescopes.**—We have thus far confined our attention to the principles of the ordinary astronomical telescope, and we have dealt with it in its simplest form. There are also Reflecting Telescopes, in which a speculum, or mirror, takes the place of the object-glass. These instruments appear in several different forms. The principle on which Herschel's is constructed, will be understood from Fig. 97.

The concave mirror SS is placed at the farthest ex-

---

lenses turned in micrometers? 432. What is the use of the telescope-tube? With what is the eye-piece furnished? 433. Where is the largest refracting telescope in the world? What is its size? How does the light received by the object-glass compare with that received by the eye? When the air is pure, how high a power does it bear? 434. What other kind of telescopes is there? In reflecting tele-

REFLECTING TELESCOPES. 239

tremity of the tube, inclined so as to make the rays that fall upon it converge toward the side

FIG. 97.—PRINCIPLE OF HERSCHEL'S REFLECTOR.

of the tube in which the eye-piece $a\,b$ is fixed to receive them. The observer at E, with his back toward the heavenly body, looks through the eye-piece, and sees the reflected image. His position is such as not to prevent the rays from entering the open end of the tube.

435. **The Largest Reflector.**—The largest reflecting telescope in the world is one constructed by the late Earl of Rosse. Its mirror is six feet in diameter, and weighs four tons. The tube at the bottom of which it is placed is fifty-two feet long and seven feet across. It is computed that, when this instrument is used, 250,000 times as much light from a heavenly body is collected as reaches the naked eye.

436. **Different Mountings.**—An astronomer uses the telescope for two kinds of work: he desires to watch the heavenly bodies, and study their physical constitution; he also wants to note their actual places and relative positions. Accordingly, he *mounts* or arranges his instrument in several different ways.

For the first kind of work the only essential is that the instrument should be so arranged as to command every portion of the sky. The best mounting for this purpose is shown in Fig. 98, which represents an eight-inch telescope equatorially mounted. With such an instrument, called an **Equatorial**, a heavenly body may be followed from its rising to its setting, the proper motion being communi-

---

scopes, what takes the place of the object-glass? Explain the principle in Herschel's reflector. 435. Give an account of the largest reflector in the world. 436. For what two kinds of work does an astronomer use the telescope? When he wants to watch a heavenly body, what alone is essential? What is the best mounting for this purpose? What is an instrument so mounted called? In

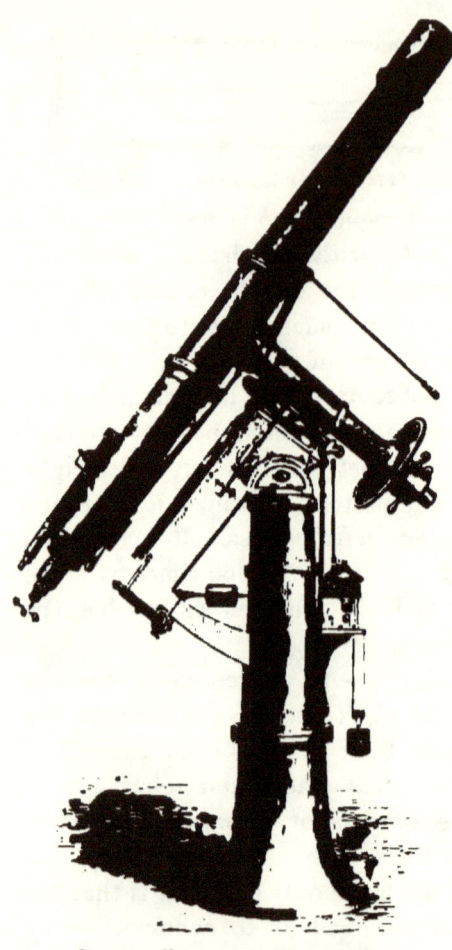

Fig. 98.—Equatorial Telescope.

cated to the instrument by clockwork.

In this arrangement, a strong iron pillar supports a head-piece, in which is fixed the *polar axis* of the instrument parallel to the axis of the Earth. This polar axis is made to turn round once in twenty-four hours by the clock shown on the right of the pillar.

It is obvious that a telescope attached to such an axis will always move in a circle of declination, and that the clock, turning the telescope in one direction as fast as the Earth is carrying it in the opposite one, will keep the instrument fixed on the object. It is inconvenient to attach the telescope directly to the polar axis, as the range is then limited; it is fixed, therefore, to a *declination axis*, placed above the polar axis and at right angles to it, as shown in Fig. 98.

---

what will a telescope thus mounted always move? How is the telescope kept

437. For the other kinds of work, telescopes are mounted as **Altazimuths, Transit-instruments, Transit-circles,** and **Zenith-sectors.**

438. **Measurement of Angles.**—In all these instruments, angles are measured by means of graduated arcs or circles attached to telescopes. The graduation is sometimes carried to the hundredth part of a second by Verniers, or small scales minutely subdivided movable by the side of larger fixed scales. It is of the greatest importance that the circle should be not only correctly graduated, but correctly *centred*—that is, that the centre of movement should be also the centre of graduation. To insure greater precision, spider-webs, or fine wires, are fixed in the focus of the telescope to point out the exact centre of the field of view. An instrument with the cross-wires perfectly adjusted, is said to be correctly *collimated*.

439. In addition to the fixed wires, movable ones are sometimes employed by which small angles may be measured. An eye-piece so arranged is called a Micrometer. The movable wire is set in a frame moved by a screw, and the distance of this wire from the fixed central one is measured by the number of revolutions and parts of a revolution of this screw, each revolution being divided into thousandths by a small circle outside the body of the micrometer.

Attached to the micrometer, or to the eye-piece which carries it, is also a **Position-circle**, divided into 360°; by this the angle made by the line joining two stars, with the direction of movement across the field of view, is determined. The use of the position-circle in double-star measurements is very important, and it is with its aid that their

---

fixed on the same object? 437. For the other kinds of work, how are telescopes mounted? 438. In all these instruments, how are angles measured? How far is the graduation sometimes carried, and how? What is of the greatest importance? To insure greater precision, what are provided? What is said of an instrument which has the cross-wires perfectly adjusted? 439. What is meant by a Micrometer? How is the movable wire fixed? What is attached to the

orbital motion has been determined. The micrometer wires, or the field of view, are illuminated at night by means of a small lamp outside, and a reflector inside, the telescope (see Fig. 99).

440. If we want simply to measure the angular distance of one celestial body from another, we use a Sextant; but, generally speaking, what is to be determined is not merely their angular distance, but their apparent position either on the sphere of observation or on the celestial sphere itself.

441. In the former case,—that is, when we wish to determine positions on the visible portion of the sky,—we use what is termed an Altitude and Azimuth Instrument, or briefly an Altazimuth. If we know the sidereal time, we can by calculation find out the right ascension and declination of a body whose altitude and azimuth on the sphere of observation we have instrumentally determined.

442. **The Altazimuth.**—An altazimuth is an instrument with a vertical central pillar supporting a horizontal axis. There are two circles; one horizontal, in which is fitted a smaller (ungraduated) circle with attached verniers fixed to the central pillar, and revolving with it; the other, vertical, at one end of the horizontal axis, and free to move in all vertical planes. To this latter the telescope is fixed. When the telescope is directed to the south point, the reading of the horizontal circle is 0°; when it is directed to the zenith, the reading of the vertical circle is 0°. Consequently, if we direct the telescope to any particular star, one circle gives the zenith distance of the star (or its altitude); the other gives its azimuth.

If we fix or *clamp* the telescope to the vertical circle, we can turn the axis which carries both round, and ob-

---

micrometer? What is the use of the Position-circle? How are the micrometer wires illuminated at night? 440. If we want simply to measure the angular distance of one celestial body from another, what do we use? Generally speaking, what else is to be determined? 441. What is used when we wish to determine positions on the visible portion of the sky? 442. Describe the altazimuth and its

THE ALTAZIMUTH. 243

serve all stars having the same altitude, and the horizontal circle will show their azimuths. If we clamp the axis to the horizontal circle, we can move the telescope so as to

FIG. 99.—PORTABLE ALTAZIMUTH.

make it travel along a vertical circle, and the circle attached to the telescope will give us the zenith-distances of the stars (or the altitude), which, in this case, will lie in two azimuths 180° apart.

Fig. 99 represents a portable altazimuth, the various parts of which will be recognized from the foregoing description.

443. **The Transit-circle.**—When we wish to determine directly the position of a heavenly body on the celestial sphere, a Transit-circle is used. This instrument consists of a telescope movable in the plane of the meridian, being supported on two pillars, east and west, by means of a horizontal axis. The ends of the axis are of exactly equal size, and move in pieces, which, from their shape, are called Y's. When the instrument is in perfect adjustment, the line of collimation of the telescope is at right angles to the axis, the axis is exactly horizontal, and its ends are due east and west. Under these conditions, the telescope describes a great circle of the heavens, passing through the north and south points and the celestial pole; that is, in all positions it points to some part of the meridian of the place.

On one side of the telescope is fixed a circle, which is read by microscopes attached to one of the supporting pillars. The cross-wires in the eye-piece of the telescope enable us to determine the exact moment of sidereal time at which the meridian is crossed; this time is the right ascension of the object. The circle attached shows us its distance from the celestial equator; this is its declination. So by one observation, if the clock is right, the instrument perfectly adjusted, and the circle correctly divided, we get both coördinates.

444. **Determination of Positions with the Transit-circle.**—As we have already seen, a celestial meridian is nothing

---

mode of operation. 443. When we wish to determine directly the position of a heavenly body on the celestial sphere, what is used? Of what does the transit-circle consist? When the instrument is perfectly adjusted, what does the telescope describe? How are right ascension and declination found with the transit

but the extension of a terrestrial one; and as the latter passes through the poles of the Earth, the former will pass through the poles of the celestial sphere: consequently, wherever we may be, the northern celestial pole will lie somewhere in the plane of our meridian. If the position of the pole were exactly marked by the pole-star, this star would remain immovable in the meridian; and when a celestial body was also in the meridian, if we adjusted the circle so as to read 0° when the telescope pointed to the pole, we could determine the north-polar distance of the body by simply pointing the telescope to it, and noting the angular distance shown by the circle.

But, as the pole-star does not lie exactly at the pole, we have to adopt some other method. We observe the zenith-distance of a circumpolar star when it passes the meridian above the pole, and also when it passes below it, and taking half the sum of these zenith-distances, we find the zenith-distance of the celestial pole. The celestial equator, which is 90° from the celestial pole, can then be readily determined; its zenith-distance will be the difference between the zenith-distance of the celestial pole, already known, and 90°. The horizon, which is 90° from the zenith, can also be determined. We can, therefore, measure angular distances with our transit-circle,

 I. From the zenith.
 II. From the celestial pole.
 III. From the celestial equator.
 IV. From the horizon.

Any of these distances can easily be turned into any other.

445. When we have obtained the distance from the celestial equator, we get in the heavens the equivalent of

---

circle? 444. If the celestial pole exactly corresponded with the polar star, how could we determine the north-polar distance of a body? As it is, how do we find the north celestial pole? What will the zenith-distance of the celestial equator be equal to? How can the horizon be determined? From what, therefore, may we measure angular distances with the transit-circle? 445. What is distance

terrestrial latitude. But this is not enough; a hundred places may have the same latitude, a hundred stars may have the same declination; we need what is called another coördinate, to fix their position. On the Earth we get this other coördinate by reckoning from the meridian which passes through the centre of the transit-circle at Greenwich. So in the heavens we reckon from the position occupied by the Sun at the vernal equinox.

The astronomer has, not only a telescope and circle, but also a sidereal clock, adjusted (as already stated) to the apparent movement of the stars, or the actual rotation of the Earth. Sidereal time, like right ascension, is reckoned from the first point of Aries. Hence, a sidereal clock at any place will denote the right ascension of the celestial meridian visible in the transit-circle at that moment; and if we at the same moment, by means of the circle, note how far a heavenly body is from the celestial equator, we shall know both its right ascension and declination, and its place in the heavens will be determined. The Earth itself, by its rotation, brings every star in turn to the meridian of our place of observation, and thus performs the most difficult part of the work for us.

446. In order that the angular distance from the zenith, and the time of meridian passage, may be correctly determined, observations of the utmost delicacy are required.

The circle of the transit used at Greenwich is read by the microscopes in six different parts of the limb at each observation, and the recorded zenith-distance is the mean of these readings. The right ascension is obtained with equal care. The transit of the star is watched over nine equidistant wires, in the micrometer eye-piece (called in

---

from the celestial equator called? What besides declination is needed, to determine a heavenly body's position? How is right ascension obtained? How does the Earth itself assist us in finding it? 446. What facts are mentioned, to show the care with which observations are made? 447. How many methods are there

this case a *transit eye-piece*), the middle one being exactly in the axis of the telescope.

447. **Methods of determining the Time of Transit over a Wire.**—There are two methods of observing the time of transit over a wire, one called the *eye and ear method*, the other the *galvanic or chronographic method*. In the former, the observer, taking his time from the sidereal clock, which is always close to the transit-circle, listens to the beats, and estimates at what interval between two beats the star passes behind each wire. An experienced observer mentally divides a second of time into ten equal parts with no great effort.

In the second method, an apparatus called a Chronograph is used. A barrel covered with paper is made to revolve at a uniform rate of speed. By means of a galvanic current, a pricker attached to the keeper of an electro-magnet is made at each beat of the sidereal clock to puncture the revolving barrel. The pricker is carried along the barrel, so that the punctures, about half an inch apart, form a spiral. Here, then, we have the flow of time fairly recorded on the barrel. At the beginning of each minute the clock fails to send the current, so that there is no confusion. What the clock does regularly at each beat, the observer does when a star crosses the wires of his transit eye-piece. He presses a spring, and an additional current at once makes a puncture on the barrel. The time at which the transit of each wire has been effected, is estimated from the position the additional puncture occupies between the punctures made by the clock at intervals of a second.

The observer is thus enabled to confine his attention to the star. After completing his observation, he can at leisure make the necessary notes on the punctured paper

---

of determining the time of transit over a wire? What are they called? Describe "the eye and ear method." What is used in the second method? Give an account of the mode of using the chronograph. What advantage has the observer

which is taken off the barrel when filled, and bound up as a permanent record.

448. Determination of Positions with the Equatorial.—With the transit-circle, the position of a body in the celestial sphere can be determined only when it is on the meridian. The equatorial enables this to be done, on the other hand, in every part of the sky, though not with such extreme precision. The object is brought to the cross-wires of the micrometer eye-piece, and the declination-circle at once shows its declination. The right ascension is determined as follows:—At the lower end of the polar axis is a movable circle divided into the 24 hours. Flush with the graduation are two verniers; the upper one fixed to the stand, the lower one movable with the telescope. The fixed vernier shows the position occupied by the telescope, and therefore by the movable vernier, when the telescope is exactly in the meridian. Prior to the observation, the circle is adjusted so that the local sidereal time—or the right ascension of the part of the celestial sphere in the meridian—is brought to the fixed vernier. The circle is then moved by the clockwork of the instrument; and when the cross-wires of the telescope are adjusted on the object, the movable vernier shows its right ascension on the same circle.

449. Star-catalogues.—The method which is good for determining the exact place of a single heavenly body is good for mapping the entire heavens; accordingly, the whole celestial sphere has been mapped out, the right ascension and declination of every object having been determined.

The most important of the catalogues in which these positions are contained, is due to the German astronomer

---

in this method? 448. As regards the determination of positions, how does the equatorial differ from the transit-circle? How is declination obtained with the equatorial? How is right ascension determined? 449. What has been accomplished through these methods of finding the declination and right ascension?

CORRECTION OF OBSERVATIONS. 249

Argelander. This catalogue contains the positions of upward of 324,000 stars, from N. Decl. 90° to S. Decl. 2°. Bessel also has put forth a catalogue of more than 32,000 stars. Airy and the British Association have published similar lists. There are also catalogues dealing with double and variable stars exclusively.

450. **Corrections to be applied.**—After the astronomer has made his observations of a heavenly body, and has freed them from instrumental and clock errors, he has obtained what is termed the *observed* or *apparent place*. This, however, is worth very little; he must, in order to obtain its *true place*, apply other corrections.

451. **Correction for Refraction.**—The first correction is needed, to nullify the effect of refraction already explained. Refraction causes a heavenly body to appear higher the nearer it is to the horizon. On an object in the zenith it has no effect: on one near the horizon, its effect is very decided; at sunset, for instance, in consequence of refraction, the Sun appears above the horizon after it has actually sunk below it.

The correction, therefore, to be made for refraction, depends entirely on the altitude of the body on the sphere of observation. Table VII. in the Appendix shows the amount of correction for different altitudes. In practice, the corrections are themselves corrected according to the density of the air at the time of observation.

452. **Correction for Aberration.**—We have already alluded to the aberration of light (Art. 410). It results from the fact that the observer's telescope, carried round by the Earth's annual motion round the Sun, must always be pointed a little in advance of the star, in order, as it were,

---

By whom have star-catalogues been published? 450. After the astronomer has found the apparent place of a heavenly body, what has he yet to do? 451. For what is the first correction needed? What is the effect of refraction on a heavenly body in different positions? On what, therefore, does the correction to be made for refraction entirely depend? In practice, according to what are the corrections themselves corrected? 452. For what is the next correction to

to catch the light from it. Hence the star's *aberration-place* will be different from its *real place;* and, as the Earth travels round the Sun, and the telescope is carried round with it always pointed ahead of the star's place, the aberration-p l a c e revolves round the real place exactly as the Earth (if its orbit be regarded as circular) would be seen from the star to revolve round the Sun. The aberration-places of all stars, in fact, describe circles parallel to the plane of the Earth's orbit. If the star lie at the pole of the ecliptic, the path of its aberration-place will appear as a circle, the centre of which will be at its true place. The aberration-place of a star in the ecliptic will oscillate backward and forward, as we are in the plane of the circle; that of one in a middle celestial latitude will appear to describe an ellipse. The diameter of the circle, the major axis of the ellipse, and the amount of oscillation, will all be equal—about 40.5″; but the minor axis of the ellipses described by the stars in middle latitudes will increase from the equator to the pole. The correction to be made is half of the above-mentioned invariable quantity, or 20.25″, which is called the *constant of aberration*. It is determined by the following proportion, bearing in mind that the 360° of the Earth's orbit are passed over in $365\frac{1}{4}$ days, and that light takes about 8 minutes 13 seconds to come from the Sun:—

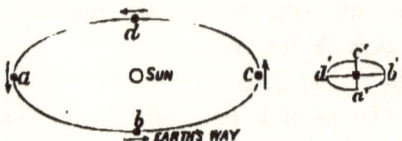

FIG. 100.—EFFECT OF ABERRATION: $a, b, c, d,$ the Earth in different parts of its orbit; $a', b', c', d',$ the corresponding aberration-places of the star, varying from the true place in the direction of the Earth's motion at the time.

$$\text{Days.} \qquad \text{m. s.} \qquad ° \qquad ″$$
$$365\tfrac{1}{4} \;:\; 8\;\;13 \;::\; 360 \;:\; 20.25$$

453. The direction of the Earth's motion in its orbit,

---

be made? From what does aberration result? How does the aberration-place move, in the case of stars in different positions? What is the allowance to be made for aberration? What is it called? How is the constant of aberration determined? 453. What is meant by the Earth's way? How far is it from the

called the *Earth's way*, referred to the ecliptic, is always 90° behind the Sun's position in the ecliptic at the time; therefore the aberration-place of a star will lie on the great circle passing through the star and the point in the ecliptic 90° behind the Sun.

454. **Correction for Parallax.**—Observations of the celestial bodies comparatively near the Earth, such as the Moon and some of the planets, when made at different places on the Earth's surface, though corrected as we have indicated, do not give the same result, as their positions on the celestial sphere appear different to observers at different points of the Earth's surface. This effect will be readily understood by changing our position with regard to any near object, and observing it as projected on different backgrounds in the landscape. The nearer we are to the object, the more will its position appear to change. To get rid of these discrepancies, the observed positions are further corrected to what they would be were the observations made at the centre of the Earth. This is called applying the correction for *parallax*.

455. **Parallax** is the angle under which a line drawn from the observer to the centre of the Earth would appear at the body observed; in other words, it is the angle formed at the body in question by two lines drawn one to the observer's eye and the other to the Earth's centre. When a body is at the zenith, it has no parallax. When it is on the horizon, its parallax, which is then termed its Horizontal Parallax, is greatest.

This is obvious from Fig. 101. C being the Earth's centre, and O an observer, a body at Z (the zenith) is seen in exactly the same direction from both points, and has no parallax. At S its parallax is OSC, and at H it is

---

Sun's position in the ecliptic? In what, therefore, will the aberration-place of a star lie? 454. Why do observations made at different parts of the Earth's surface have to be corrected to what they would be if made from the Earth's centre? What is this correction called? 455. What is Parallax? What is Horizontal Parallax? Where is parallax least, and where greatest? Show this with Fig.

252   ASTRONOMICAL INSTRUMENTS.

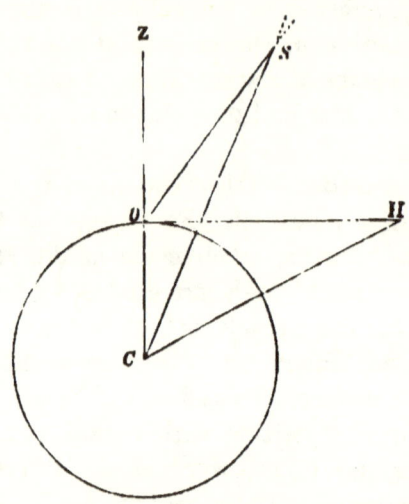

FIG. 101.—PARALLAX.

$O H C$, which is greater than $O S C$ or the angle that would be formed at any point between S and H.

As shown by the dotted prolongations of the lines $O S$, $C S$, a body seen from O would appear farther from Z than if seen from C; hence, to obtain the true zenith-distance, we must subtract the correction for parallax from the apparent zenith-distance.

456. Changes in Positions already determined.—We have seen that the positions of the heavenly bodies are determined with reference to either the plane of the ecliptic or the plane of the equator; and that from one of the points of intersection of these two planes—that, namely, occupied by the Sun at the vernal equinox, called the first point of Aries and written ♈—right ascension and celestial longitude are both reckoned. If these planes, then, are changeless, a position once determined will be determined forever; but if either plane varies, then the point of intersection will of course vary, and corrections in the positions of the stars as once determined will be necessary from time to time. Now, it is found that changes occur in both planes.

457. It has been stated that the Earth's axis always points in the same direction. Strictly speaking, this is

---

101. How must the correction for parallax be used, to obtain the true zenith-distance? 456. What renders corrections in the positions of the stars, as once determined, necessary from time to time? 457. What change takes place in the position of the Earth's pole? From this what important fact follows? How does

not the case. The pole of the Earth is constantly changing its position, and revolves round the pole of the ecliptic in 25,868 years, so that the pole-star of to-day will not be the pole-star 3,000 years hence.

From this a very important fact follows. As the Earth's axis changes, the plane of the equator changes with it, and so that each succeeding vernal equinox happens a little earlier than it would otherwise do. This is called the **Precession of the Equinoxes**, because the equinox seems to move backward, or from left to right, so as to meet the Sun earlier. In the time of Hipparchus—2,000 years ago —the Sun at the vernal equinox was in the constellation Aries; it is now in the constellation Pisces.

458. The plane of the ecliptic is also subject to variation. This is termed the **Secular Variation of the Obliquity of the Ecliptic.**

459. Of these changes, the precession of the equinoxes is the more important. It causes the point of intersection of the two fundamental planes to recede 50.37572″ annually. To this is due the difference in length between the sidereal and the tropical year.

460. **Cause and Effect of these Changes.**—The cause of these changes is the attraction exercised by the Sun, Moon, and planets, upon the protuberant equatorial portions of our Earth. The effect is to render both latitudes and longitudes, right ascensions and declinations, variable. Hence the observed position of a heavenly body to-day will not be the position occupied last year, or to be occupied next year. Apparent positions have to be corrected, to bring them to some common epoch, such as 1850, 1880, etc., so that they may be strictly comparable.

---

the equinox seem to move? What is this motion called? Since the time of Hipparchus, what change has taken place in the position of the Sun at the vernal equinox? 458. What is the variation in the plane of the ecliptic called? 459. Of these changes, which is the more important? What is the amount of recession annually? What does this recession cause? 460. What is the cause of these changes in the plane of the ecliptic and the plane of the equator? What is their

461. Celestial Latitude and Longitude, how determined.
—Celestial latitude and longitude, which are used to determine the position of heavenly bodies with reference to the ecliptic, are not obtained by observation, but are calculated from the true right ascension and declination by means of spherical trigonometry.

462. Recapitulation.—Let us recapitulate what has been said as to the methods by which the true positions of the heavenly bodies are obtained:—

1. The astronomer, to make observations on that part of the celestial sphere which is visible to him, makes use principally of a sextant or an altazimuth. The positions of a body thus determined may by calculation be referred to the celestial sphere itself, and its right ascension and declination determined.

2. Observations of a body with regard to the celestial sphere itself are made principally by means of a transit-circle or an equatorial, by which both apparent right ascension and declination may be directly determined.

3. In all observations, the instrumental and clock errors are carefully corrected.

4. Besides the instrumental and clock errors, there are others—caused by the refraction and aberration of light,—which must also be corrected.

5. Besides these, another error, parallax, results from the observer's position on the Earth's surface. This is corrected by reducing all observations to the Earth's *centre*.

6. There are still other errors depending upon the change of the intersection of the two planes to which all measurements are referred. These are got rid of by reducing all observations to a point of time (as parallax was got rid of by reducing them to a point of space—the centre

of the Earth). Some year is fixed upon, and the observations reduced to what they would have been at this time if the year is past, or what they will be when made at this time if the year is to come.

7. The right ascension and declination are easily converted by calculation into celestial longitude and latitude, if required.

463. By means of observations freed from these errors, and extending over centuries, astronomers have been able to determine the positions of all the stars with the greatest accuracy, and to discover the proper motions of some of their number. They have also investigated the motions of the bodies of our system so thoroughly as to ascertain the laws by which they are regulated, and to be able to predict their exact positions for years to come.

This information is embodied in an Almanac or Ephemeris, published in advance by each of the principal Governments, for the use of travellers and navigators. The United States and the English Nautical Almanac are publications of this character, in which are given, with most minute accuracy, the positions of the principal stars, the planets, and the Sun, from day to day, and the positions of the Moon from hour to hour. These positions enable us to determine—I. Time. II. Latitude. III. Longitude.

Determination of Time, Latitude, and Longitude.

464. **Determination of Time.**—When time only is required, a transit-instrument is used; that is, a simple telescope mounted like the transit-circle, *but without the circle*, or with only a small one—the transits of stars, the right ascension of which has been already determined with great accuracy by transit-circles, in fixed observatories, being

---

obtained. 463. By means of observations thus freed from errors and extending over centuries, what have astronomers been able to do? In what is this information embodied? What are given in the Nautical Almanac? What do these positions enable us to determine? 464. When time only is required, what is used? How

observed. This gives us the local sidereal time, which may, if necessary, be converted into mean solar time.

465. **Determination of Latitude.**—To determine our position on the Earth's surface, all we need is our latitude and longitude. The determination of the former in a fixed observatory is an easy matter, if proper instruments be at hand. For instance: half the sum of the altitudes (corrected for refraction) of a circumpolar star, at upper and lower culmination, even if its position is unknown, will give us the elevation of the pole, and therefore the latitude of the place.

Again, if we find the zenith-distance of a star, the declination of which has been accurately determined, we can readily obtain the latitude. For, as declination is referred to the plane of the terrestrial equator prolonged to the stars, it is the exact equivalent of terrestrial latitude. If a star of 0° declination is observed exactly in the zenith, the observer must be on the equator; if the declination of a star in the zenith is 45°, then our latitude is 45°; if a star of declination 39° N. passes 10″ to the north of our zenith, then our latitude is 38° 59′ 50″, and so on.

466. On board ship, and in the case of explorers, the problem is for the most part limited to determining the meridian altitude of the Sun or Moon, as the sextant only can be employed. Suppose such an observation to give the altitude as 29° from the south point of the horizon—equivalent to 61° zenith-distance—and that the Nautical Almanac gives its declination on that day as 12° south; ir we were in lat. 12° S. the Sun would be overhead, and its zenith-distance would be 0°; as it is 61° to the south, we are 61° to the north of 12° S., or in N. lat. 49°. So, if we find the meridian altitude to be 10° from the north point of the horizon (or the zenith-distance to be 80°), and the

---

is the local sidereal time obtained? 465. What do we need, to determine our position on the Earth's surface? In what two ways can we find the latitude in a fixed observatory? Illustrate the latter method. 466. How can latitude be deter-

Nautical Almanac gives the declination at the time as 20° N., our position will be in 60° S. lat.

467. Determination of Longitude.—Longitude is in fact *time*, and difference of longitude is the difference of the times at which the Sun crosses any two meridians, the twenty-four hours solar mean time being distributed among the 360° of longitude, so that 1 hour = 15°, and so on.

Several ways of determining longitude are employed in fixed observatories. The most convenient one consists in electrically connecting the two stations whose difference of longitude is sought, and observing the transit of the same stars at each. Thus the transits at station A are recorded on the chronograph at stations A and B, and the transits at station B are similarly recorded at B and A; from both chronographs the interval between the times of transit is accurately recorded in sidereal time, and the mean of all the differences converted into mean solar time gives the difference of longitude.

468. One mode of determining longitude at sea, which consists in finding the difference between local time, and Greenwich time as indicated by an accurate chronometer, and converting this difference into difference of longitude, has been explained in Art. 191.

A second method consists in making use of the heavens as a dial-plate, and of the Moon as the hand. In the Nautical Almanac, the distances of the Moon from the stars in her course are given, as they would appear if observed from the Earth's centre, for every third hour in Greenwich time. The sailor, therefore, observes the Moon's distance from the stars in question, and corrects his observation for refraction and parallax. Referring to the Nautical Almanac, he sees the time at Greenwich at which

---

mined at sea? Give examples. 467. To what is longitude, and to what is difference of longitude, really equivalent? What is the most convenient mode of determining longitude in fixed observatories? 468. What method of determining longitude at sea has already been explained? What other method is there?

the distance is the same as that which he has obtained; and knowing the local time (from day observations) at the instant at which his observation was made, he readily finds the difference of time, and thence the difference of longitude.

Determination of Distances.

469. To determine the distances of the heavenly bodies, astronomers have recourse to methods similar to those used by surveyors in measuring distances on the Earth. When two angles of a triangle and the length of the included side are known, the remaining angle and sides can with the aid of trigonometry be determined. Accordingly, a base-line which subtends an appreciable angle at the body in question, and whose length is accurately found, being taken, a triangle is formed by drawing lines from the ends of the base to the object. The angles at the extremities of the base-line being then determined by observation, the parallax of the body and its distance from the Earth can be found.

470. **Parallax of the Moon.**—In the case of the Moon, the base-line taken is the distance between two places on the Earth's surface remote from each other, which distance can be determined from their positions on the globe when the size of the Earth is known. The mean equatorial horizontal parallax of the Moon has thus been found to be nearly 57′ 3″.

471. **Determination of the Distance of Mars.**—In the case of Mars, the Earth's diameter is made the base-line, observations being taken at the same place at an interval of 12 hours, which owing to the Earth's rotation separates the points of observation by the length of the Earth's

---

469. To determine the distances of the heavenly bodies, to what do astronomers have recourse? Explain the mode of proceeding. 470. What is taken for a base-line in the case of the Moon? What is the mean equatorial horizontal parallax of the Moon found to be? 471. What is made the base-line in the case of Mars?

diameter—allowance being made for the motion of both planets in the interval between the observations.

472. The Sun's Parallax.—As seen from the Sun, the Earth's diameter is so small that it is useless as a base-line in determining the Sun's distance. This can, however, be obtained directly by a method pointed out by Halley in 1716, based on observations of the transit of Venus. Unfortunately, these transits happen but rarely; the last took place in 1874, the next available one will be in 1882. On the other hand, when they do occur, as the planet is projected on the Sun, the Sun serves the purpose of a micrometer, and observations may be made with the most rigorous exactness.

The old value of the Sun's parallax, obtained by Bessel from the transit of Venus, was		8.578″
New value obtained by Hansen from the Moon's parallactic equation,		8.916″
"	" Winnecke from observations of Mars,	8.964″
"	" Stone,	8.930″
"	" Foucault, from the velocity of light,	8.960″
"	" Le Verrier, from the motions of Mars, Venus, and the Moon,	8.950″

The difference between the old and the new value now generally accepted, which equals about two-fifths of a second of arc, amounts to no more than the apparent breadth of a human hair viewed at the distance of about 125 feet. Yet it requires us to alter the distance and diameter of nearly every body in the solar system, and makes a differ-

---

How is this done? 472. On what is the method of finding the Sun's parallax based? Why is not the Earth's diameter used as a base-line? When will the next transit of Venus occur? What was the old value of the Sun's parallax, obtained from the transit of Venus? What later values have been obtained? What is the difference between the old and the new value now generally accepted?

ence of about 3½ millions of miles in the distance of the Sun itself. According to the old value of the parallax, the Sun's distance was about 95,000,000 miles; according to the new, it is but 91,430,000. The transit of 1874 was observed with the greatest care, to see whether this new value of the solar parallax would be confirmed.

473. **Parallax of the Stars.**—Having thus obtained the distance of the Sun, we have a base-line of enormous dimensions; for the positions successively occupied by our Earth in two opposite points of its orbit will be 183,000,000 miles apart, and we can make this our base-line by taking observations at the same place at an interval of six months. But we find that even this great line is insufficient to measure the distances of the stars. In almost every case, there is no apparent difference in the position as observed in January and July, February and August, etc. As seen from the fixed stars, the Earth's orbit is but a point!

Now, an instrument such as is ordinarily used should show us a parallax of one second—that is, an angle of 1″ subtended at the star by half the base-line we are using; and a parallax of 1″ means that the object is 206,265 times farther away than we are from the Sun, as the Sun's distance is the half of our base-line. If, then, a star's parallax be less than 1″, the star must be farther away than 206,265 times 91,430,000 miles!—and this we find to be the case with every star in the heavens.

474. In the great majority of cases, the true zenith-distance of a star is the same all the year round. As this true place results from the several corrections referred to in Art. 462, even when there is a slight variation, it may be wrong to ascribe it to parallax. A slight error in the

---

What difference in the Sun's distance does this small difference of parallax make? 473. What may now be taken as a base-line, to find the parallax of the stars? How may it be made available? Is it found sufficient for the purpose? Why not? How great a parallax should an ordinary telescope show us? If a star's parallax be less than one second, how far must the star be away? What follows with respect to the distance of every star in the heavens? 474. In the case of what star alone was the parallax found by this method? What method was

PARALLAXES OF THE STARS.

refraction, or the presence of proper motion in the star, would give rise to a greater difference of position than the one due to parallax, as in no case does the latter exceed 1″. Hence, as long as the problem was approached in this manner, very little progress was made, the parallax of *a Centauri* (0.9187″) alone being obtained.

Bessel, however, employed a method by which the various corrections were done away with, or nearly so. He chose a star having a decided proper motion, and compared its position, night after night, by means of the micrometer only, with other small stars lying near it which had no proper motion, and which therefore he assumed to be very much farther away. He found that the star with the proper motion did really change its position with regard to the more remote ones, as it was observed from different parts of the Earth's orbit. This method has since been pursued with great success. Here is a table showing the parallax and distance of some of the nearer stars, as obtained by this method.

Star.	Parallax.	Distance. Sun's distance = 1.
a Centauri . .	0.9187″	224,000
61 Cygni	0.5638	366,000
1830 Groombridge .	0.226	912,000
70 Ophiuchi . . .	0.16	1,286,000
Vega	0.155	1,337,000
Sirius	0.15	1,375,000
Arcturus	0.127	1,624,000
Polaris	0.067	3,078,000
Capella	0.046	4,484,000

employed by Bessel? How did it succeed? What star is nearest to the Earth? What is its parallax, and what its distance in terms of the Sun's distance? What is the next nearest star? Give the parallax and distance of 61 *Cygni*. Of

Thus *a Centauri*, which is the nearest star, is found to be 224,000 times as far off as the Sun, or more than 20,000,000,000,000 miles.

Determination of the Size of the Heavenly Bodies.

475. When the distance of a body is known, and also its angular measurement, its size is determined by a simple proportion; for the distance is, in fact, the radius of the circle on which the angle is measured.

There are 1,296,000 ($360 \times 60 \times 60$) seconds in an entire circumference. Hence, as the circumference is 3.1416 times the diameter, and the diameter is twice the radius, there are as many seconds in that part of the circumference which equals the radius as twice 3.1416 is contained times in 1,296,000—or 206,265″. Hence the following proportion :—

$$\left\{\begin{array}{c}\text{The diameter}\\ \text{of the body}\\ \text{in miles}\end{array}\right\} : \left\{\begin{array}{c}\text{the distance}\\ \text{in miles}\end{array}\right\} :: \left\{\begin{array}{c}\text{the angular}\\ \text{diameter}\\ \text{in seconds.}\end{array}\right\} : \left\{206265''\right\}$$

Calling the diameter in miles d, by multiplying the means together and dividing their product by the given extreme, we get the following formula :—

$$d = \frac{\text{distance} \times \text{angular diameter}}{206265} \quad \ldots \quad (1)$$

The mean angular diameter of the Moon is 31′ 8.8″, or 1868.8″; its distance is 237,640 miles; what is its diameter in miles?

Applying Formula 1, we have

$$d = \frac{237640 \times 1868.8}{206265} = 2153 \text{ miles.}$$

In Table II. of the Appendix are given the greatest and least apparent angular diameters of the planets, as seen from the Earth. Hence the mean angular diameters can be found, and with these and the distances given in Art. 367, the student can calculate the real diameters for himself.

Vega. Of Sirius. Of Arcturus. 475. From what can the size of a heavenly body be determined, and how? Give the process by which the formula for finding the diameter in miles can be obtained. Give the formula. Apply this formula, to

476. From Formula 1 we derive Formula 2 given below, which is to be used when the diameter in miles and the angular diameter are known, and the distance is required:—

$$\text{Distance} = \frac{206265 \times d}{\text{angular diameter in seconds}} \quad . \quad . \quad (2)$$

The diameter of the Sun being 852,584 miles, and its mean angular diameter 32' 4.205", what is its distance?

CHAPTER XV.

THE SPECTRUM.

477. A CAREFUL examination of the solar spectrum has revealed to us the importance of solar radiation (Art. 127). Not only may we liken the gloriously-colored bands which we call the spectrum to the key-board of an organ—each ray a note, each variation in color a variation in pitch—but as there are sounds in nature which we cannot hear, so there are rays in the sunbeam which we cannot see.

What we do see is a band of color extending from red, through orange, yellow, green, blue, and indigo, to violet; but at either end the spectrum is continued. There are dark rays before we get to the red, and other dark rays after we leave the violet—the former heat rays, the latter chemical rays. This accounts for the threefold action of the sunbeam; heating power, lighting power, and chemical power.

478. **Gradual Formation of a Spectrum.**—When a cool body, such as a poker, is heated in the fire, the rays it

---

find the diameter of the Moon. 476. Give the formula for finding the distance, when the diameter in miles and the angular diameter are known. Apply this formula, to find the distance of the Sun.

477. What has been revealed to us by an examination of the solar spectrum? What do we see in the spectrum? What are there that we do not see? What three kinds of rays are combined in the sunbeam? 478. Give an account of the

first emits are invisible; if we look at it through a prism, we see nothing, though we easily perceive by the hand that it is radiating heat. As it is more highly heated, the radiation gradually increases, until the poker becomes of a dull-red color, the first sign of incandescence; in addition to the dark rays it previously emitted, it now sends forth waves of red light, which a prism will show at the red end of the spectrum. If we still increase the heat and continue to look through the prism, we find, added to the red, orange, then yellow, then green, then blue, indigo, and violet; when the poker is white-hot, all the colors of the spectrum are present. If, after this point has been reached, the substance allows of still greater heating, it will give out with increasing intensity the rays beyond the violet, until the glowing body can rapidly act in forming chemical combinations, a process which requires rays of the highest refrangibility—the so-called chemical, actinic, or ultra-violet rays.

479. **Fraunhofer's Lines.**—We owe the discovery of the prismatic spectrum to Sir Isaac Newton, but the beautiful coloring is but one part of it. Dr. Wollaston, in 1802, discovered that there were dark lines crossing the spectrum in different places. These have been called Fraunhofer's Lines, as an eminent German optician of that name afterward mapped the plainest of them with great care; he also discovered that there were similar lines in the spectra of the stars. The explanation of these dark lines we owe mainly to Kirchhoff. The law which explains them was, however, first proved by Balfour Stewart.

480. **Experiments with the Spectroscope.**—We shall observe the lines best if we make our sunbeam pass through an instrument called a **Spectroscope**, in which

---

successive steps in the formation of a spectrum by a body subjected to heat. 479. By whom was the prismatic spectrum discovered? What discovery was made by Wollaston? What are these lines called, and why? Who first explained them? 480. How can we best observe the lines? Viewed with a spectroscope,

several prisms are carefully mounted. We find the spectrum crossed at right angles to its length by numerous dark lines—gaps—which we may compare to silent notes on an organ. Now, if we light a match and observe its spectrum, we find that it is *continuous;* it runs from red through the whole gamut of color, to the visible limit of the violet; there are no gaps, no dark lines, breaking up the band.

Another experiment. Let us burn something which does not burn white; some of the metals will answer our purpose. We see at once by the brilliant colors that fall upon our eye from the vivid flame that we have here something different. The spectrum, instead of being continuous as before, now consists of two or three lines of light in different parts; as if on an organ, instead of pressing down all the keys, we sounded but one or two notes in the bass, tenor, or treble.

Let us try still another experiment. We will so arrange our prism, that while a sunbeam is decomposed by its upper portion, a beam proceeding from burning sodium, iron, nickel, copper, or zinc, may be decomposed by the lower one. We shall find in each case, that when the *bright lines* of which the spectrum of the metal consists flash before our eyes, they will occupy exactly the same positions in the lower spectrum as some of the *dark bands* do in the upper solar one.

481. Here, then, is the germ of Kirchhoff's discovery, on which his hypothesis of the physical constitution of the Sun is based; here is the secret of the recent additions to our knowledge of the stars, for stars are suns.

Vapors of metals, and gases, absorb those rays which the same vapors of metals and gases themselves emit.

---

what appearances does the solar spectrum present? Describe the spectrum of a match. Describe the spectrum of a substance that does not burn with white light—such as some of the metals. Give an account of the third experiment with the spectroscope. 481. What principle is at the basis of Kirchhoff's hypothesis?

482. Facts established by Experiment.—By experimenting in this manner, the following facts have been established:—

I. When solid or liquid bodies are incandescent, they give out continuous spectra.
II. When any gas, or solid or liquid body reduced to the state of gas, burns, the spectrum consists of bright lines only, and these lines are different for different substances.
III. When light from a solid or liquid passes through a gas, the gas absorbs those particular rays of which its own spectrum consists.

483. Fraunhofer's Lines explained.—We now see what has become of those rays which the dark lines in the solar spectrum tell us are wanting. Before they left the regions of our incandescent Sun, *they were arrested by those particular metallic vapors and gases in his atmosphere with which they beat in unison;* and the assertion that this and that metal exists in a state of vapor in the Sun's atmosphere, is based upon their absence. So various and constant are the positions of the bright bands in the spectra we can observe here, and so entirely do they correspond with certain dark bands of the spectrum of the Sun, that it has been affirmed that the chances for the correctness of the hypothesis are something like 300,000,000 to 1.

484. Spectra of the Stars.—Fraunhofer was the first to apply this discovery to the stars; and we have lately reaped a rich harvest of facts, in the actual mapping down of the spectra of several of the brightest stars, and the examination, more or less cursory, of a very large number. In

---

482. What three facts as to spectra have been established by experiment? 483. In view of these facts, how are Fraunhofer's lines explained? What do we conclude from the absence of certain rays in the spectrum? 484. To what has this discovery been extended? What is found in the case of every star whose spec-

every case, we find an atmosphere sifting out the rays that beat in unison with the metallic and gaseous vapors which it contains, and sending to us the residuum, a broken spectrum abounding in dark spaces.

485. **Importance of these Researches.**—A few words will show the great importance of these facts. They tell us that, as the solar spectrum contains dark lines, the light is due to solid or liquid particles in a state of great heat, or *incandescence;* and that the light given out by these particles is sifted, so to speak, by its atmosphere, which consists of the vapors of the substances incandescent in the photosphere. Further, as the lines in the reversed spectra occupy the same positions as the bright lines given out by the glowing particles would do, and as we can by experimenting on the different metals *match* many of the lines exactly, we can thus see which light is abstracted, and what substance gives out this light. Having done this, we know what substances (Art. 126) are burning in the Sun.

Again, we find that all the stars are more or less like the Sun, for their spectra exhibit nearly the same appearances; we can also tell, as above, what substances are burning on their surfaces (Art. 83).

486. **Spectra of the Nebulæ.**—The spectra of the nebulæ, instead of resembling that of the Sun and stars,—that is, showing a band of color with black lines across it,—consist of a few bright lines merely.

487. On August 29th, 1864, Mr. Huggins directed his telescope, armed with the spectrum-apparatus, to the planetary nebula in Draco. At first he suspected that some derangement of the instrument had taken place, for no spectrum was seen, but only a short line of light, perpen-

---

trum is examined? 485. What conclusions are drawn respecting the Sun from investigations of its spectrum? What do we find with respect to the stars? 486. Describe the spectra of the nebulæ. 487. Who examined the spectrum of the planetary nebula in Draco? Give the results of his examination. With what

dicular to the direction of dispersion. He found that the light of this nebula, unlike every other celestial light which had yet been subjected to prismatic analysis, was not composed of rays of different refrangibility, as in the case of the Sun and stars, and that therefore it could not form a spectrum. A great part of the light from this nebula consists of but one color, and was seen in the spectroscope as a bright line. A more careful examination showed another line, narrower and much fainter, a little more refrangible than the brightest line, and separated from it by a dark interval. Beyond this again, at about three times the distance of the second line, a third exceedingly faint line was seen.

The strongest line coincides in position with the brightest of the air-lines. This line is due to nitrogen, and occurs in the solar spectrum about midway between b and f, (see Frontispiece). The faintest of the lines of the nebula coincides with the line of hydrogen, marked f in the solar spectrum. The other bright line was a little less refrangible than the strong line of barium.

488. Here, then, we have three little lines forever disposing of the notion that all nebulæ are clusters of stars. With what trumpet-tongue does such a fact speak of the resources of modern science! That nebulæ are masses of glowing gas is shown by the fact that their light consists merely of a few bright lines.

An object-glass collects a beam of light which would otherwise have bathed the Earth forever invisibly to mortal eye. The beam is passed through a prism, and in a moment we find that we have no longer to do with glowing Suns enveloped in atmospheres enforcing tribute from the rays which pass through them, but with something

---

does the strongest line of this spectrum coincide? With what, the faintest line? With what does the other bright line nearly correspond? 488. What important fact respecting the physical constitution of nebulæ is established by these three little lines? 489. Describe the spectrum of the Moon. What may be inferred

devoid of atmosphere, and that something a glowing mass of gas (Art. 102).

489. Spectra of the Moon and Planets.—That moonshine is but sunshine second-hand, and that the Moon has no sensible atmosphere, is proved by the fact that in the spectroscope there is no difference, except in brilliancy, between the two. That the planets have atmospheres is shown in like manner, since in their light we find the same lines as in the solar spectrum, with the addition of other lines due to the absorption of their atmospheres.

490. Explanation of the Frontispiece.—In the Frontispiece are given representations of the solar spectrum, two stellar spectra, the spectra of the Nebula 37, II. iv., and the double line of sodium. The latter is shown, to explain the coincidences on which our knowledge of the substances present in the atmospheres of the Sun and stars depends. The light given out by the vapor of sodium consists only of the double line shown in the plate. A black double line is seen in exactly the same position in the spectra of the Sun, Aldebaran, and α Orionis; hence we infer that sodium is present in the atmospheres of all these suns.

Similarly, were we to observe the spectrum of the vapor of iron, in the same position as the 400 or 500 bright bands visible in this case, we should see coincident black lines in the spectrum of the Sun. The feeble light of the stars does not permit all these lines to be observed. It is seen in the plate that one of the bright bands in the spectrum of the nebula is coincident with one of the lines of nitrogen, and one with the hydrogen line.

491. In the spectrum of α Orionis, among eighty lines observed and measured, no less than five cases of coincidence have been detected; that is to say, we have now evidence—universally accepted in the analogous case of

---

from this? Describe the spectrum of the planets. What follows? 490. What are represented in the Frontispiece? Show how we find the substances present in the atmospheres of the stars, by taking sodium and iron as examples

the Sun—that sodium, magnesium, calcium, iron, and bismuth, are present in the atmosphere of *a Orionis*.

492. **The Spectroscope.**—The Star Spectroscope, with which these spectra have been observed, is attached to the eye end of an equatorial. As the spectrum of the *point* which the star forms at the focus is a *line*, the first thing done in the arrangement adopted is to turn this line into a band, in order that the lines or breaks in the light may be rendered visible.

The other parts of the arrangement are as follows:— A plano-convex cylindrical lens, of about fourteen inches' focal length, is placed with its axial direction at right angles to the direction of the slit, and at such a distance before the slit, within the converging pencils from the object-glass, as to give exactly the necessary breadth to the spectrum. Behind the slit, at a distance equal to its focal length, is an achromatic lens of $4\frac{1}{2}$ inches' focal length. The dispersing portion of the apparatus consists of two prisms of dense flint-glass, each having a refracting angle of 60°.

The spectrum is viewed through a small achromatic telescope, provided with proper adjustments, and carried about a centre adjusted to the position of the prisms by a fine micrometer screw. This measures to about $\frac{1}{1000}$ of the interval between A and H of the solar spectrum. A small mirror attached to the instrument receives the light which is to be compared directly with the star-spectrum, and reflects it upon a small prism placed in front of one half of the slit. This light is usually obtained from the induction-spark taken between electrodes of different metals, raised to incandescence by the passage of an induced electric current.

---

491. What is shown by an examination of the spectrum of *a Orionis*? 492. To what is the star-spectroscope attached? In the arrangement adopted, what is the first thing done? Describe the other parts of the arrangement. Through what is the spectrum viewed? How is the light which is to be compared with the star-spectrum received? How is this light usually obtained? 493. Describe the

493. A very powerful Spectroscope was for some time used at the Kew Observatory, in England, for mapping the solar spectrum. The light enters at a narrow slit in one of the collimators, which is furnished with an object-glass at the end next the prism, to render the rays parallel before they enter the prisms. In the passage through the prisms the ray is bent into a circle, widening out as it goes.

494. It is often convenient to use what is termed a Direct-vision Spectroscope—that is, one in which the light enters and leaves the prisms in the same straight line. How this is managed in the Herschel-Browning spectroscope, one of the best of its kind, by means of successive refractions and reflections, may be gathered from Fig. 102.

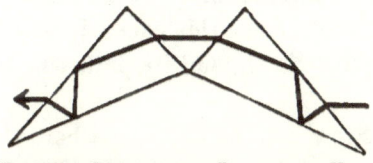

Fig. 102.—Path of the Ray in the Herschel-Browning Spectroscope.

495. **Celestial Photography.**—In both telescopic and spectroscopic observations, the visible rays of light are used. The chemical rays, however, being also present, photographs of the brighter celestial objects can be taken; and celestial photography, in the hands of Mr. De La Rue and Mr. Rutherford, has been brought to a high state of perfection. The method adopted is to place a sensitive plate in the focus of a reflector, or refractor properly corrected for the actinic rays, and then to enlarge this picture to the size required. De La Rue's photographs of the Moon, some 1¼ inches in diameter, are of such perfection that they bear subsequent enlargement to 3 feet. These pictures are now being used as a basis of a map of the Moon, 200 inches in diameter.

---

arrangement for the light in a powerful spectroscope used at Kew for examining the solar spectrum. 494. What is it often convenient to use? 495. What is said of celestial photography? What is the method adopted? What use is being made of De La Rue's photographs of the Moon?

CHAPTER XVI.

UNIVERSAL GRAVITATION.

496. **Motion.**—If a body at rest receive an impulse in any direction, it will move in that direction, and with a uniform velocity, if it be not stopped. If we set a body in motion on the Earth's surface, it will soon be stopped by friction. If we fire a cannon-ball in the air, it will in time be arrested by the resistance of the air; moreover, while its speed is slackening from this cause, it will fall, like every thing else, to the Earth, and its path will be a curved line.

Were it possible to fire a cannon in space where there is no air to resist, and were there no body to draw the ball to itself, as the Earth does, the projectile would forever pursue a straight path, with a uniform velocity. As it is, the moment the ball leaves the cannon, there is superadded to the original velocity of projection an acceleration directed toward the Earth; and the path described is what is called a *resultant* of these two velocities.

497. **Parallelogram of Forces.**—To illustrate resultant motion, suppose that the cricket-ball A, in Fig. 103, receives an impulse which will send it to B in a certain time; it will move in the direction AB. Suppose, again, it receives an impulse that will send it to C in the same time; it will move in the direction AC, and more slowly, as it has a less distance

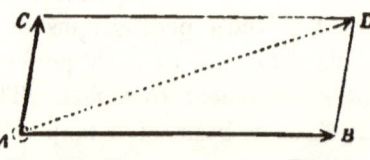

Fig. 103.—Parallelogram of Forces.

---

496. If a body at rest receive an impulse in any direction, how will it move? What soon stops a body set in motion on the Earth's surface? By what is a cannon-ball fired in the air stopped, and what is its course? Of what is the path described by such a projectile the resultant? 497. Illustrate resultant motion

to go. But suppose, again, that both these impulses are given at the same moment; it will go neither to B nor to C, but will move in a direction between these points. The exact direction, and the distance it will go, are determined by completing the parallelogram $A\,B\,C\,D$, and drawing the diagonal $A\,D$, which represents the direction and amount of the resultant motion.

498. **Weight.**—All bodies left unsupported fall to the Earth; and it is from this tendency that we derive our idea of *weight*, and of the difference between a light body and a heavy one. On the latter point, however, we must not allow the action of the atmosphere to mislead us. If we drop a dime and a feather, the latter will require more time to fall than the former; it would, therefore, at first appear that the tendency to fall, or *gravity*, of the feather is different from that of the dime. This, however, is not the case; for, if we drop them in a long tube, exhausted of air, we find that both fall in the same time. The difference in the time of falling in the air is due simply to the unequal resistance which the air offers to the bodies in their descent.

499. **Velocity of Falling Bodies.**—Machines have been invented for determining the exact rate at which a body falls near the Earth's surface. Experiments with these show that in the first second it will fall $16\frac{1}{12}$ feet, and that the velocity keeps increasing in each subsequent second. The following rules have been established:—

I. To find the space passed through during any second, multiply $16\frac{1}{12}$ feet by that one in the series of odd numbers (1, 3, 5, 7, 9, 11, etc.) which corresponds with the given second.

---

with Fig. 103. 498. Whence do we derive our idea of weight? When a dime and a feather are dropped, what do we find as regards their respective times of falling? What misapprehension might follow? How is this proved to be a misapprehension? Why does the feather take longer to fall than the dime? 499. What do experiments show with respect to the velocity of a falling body? Give the rule for finding the space passed through during any second. Give the rule for

II. To find the velocity at the termination of any second, multiply $16\frac{1}{12}$ feet by that one in the series of even numbers (2, 4, 6, 8, 10, 12, etc.) which corresponds with the given second.

III. To find the whole space passed through in any number of seconds, multiply $16\frac{1}{12}$ feet by the square of the number denoting the seconds.

Examples.—What distance will a body fall in the fourth second of its descent, what will be its velocity at the end of the fourth second, and how far will it have fallen in the first four seconds?

7 being the fourth in the series of odd numbers, in the fourth second it will fall through 7 times $16\frac{1}{12}$ feet, or $112\frac{7}{12}$ feet.

8 being the fourth in the series of even numbers, its velocity at the end of the fourth second will be 8 times $16\frac{1}{12}$ feet, or $128\frac{2}{3}$ feet, per second.

It will have fallen during the first four seconds 16 (4^2) times $16\frac{1}{12}$ feet, or $257\frac{1}{3}$ feet.

500. **Curvilinear Motion, how produced.**—If a cannon-ball were left unsupported at the mouth of a gun, it would fall to the Earth in a certain time; when fired from the gun, it has superadded to its tendency to fall a motion which carries it to the target. But during its flight gravity is constantly at work, and the law referred to in Art. 497 holds good in this case also, which is one of curvilinear motion. As the cannon-ball is pulled down from its straight course toward the target by the action of the Earth upon it, so in all cases of curvilinear motion there is a something deflecting the moving body from the rectilinear course.

501. **Newton's Discovery.**—Sir Isaac Newton was the first to see that the curved path of the Moon is similar to that of a projectile, and that both are due to the same cause as the fall of an apple—namely, the *attraction of the Earth.*

---

finding the velocity at the termination of any second. Give the rule for finding the whole space passed through in any number of seconds. Illustrate these rules with an example. 500. How is curvilinear motion produced? Show this in the case of a cannon-ball. 501. What great discovery was made by Newton? By

He saw that on the Earth's surface the tendency of bodies to fall was universal; that the Earth acted, as it were, like a magnet, drawing to itself every thing free to move, even on the highest mountains; why not, then, at the distance of the Moon? He immediately applied the knowledge derived from observations on falling bodies on the Earth, to test the correctness of his idea.

502. LAW OF GRAVITY.—Gravity is common to all kinds of matter. Its law of action may be stated thus:— *The force with which two material particles respectively attract each other is directly proportional to their masses, and inversely proportional to the square of the distances between their centres.* Now, the intensity of a force is measured by the *momentum*, or joint product of velocity and mass, produced in one second in a body subjected to its action,—and this measure of force must be remembered in discussing the above law of gravity.

Thus, if our unit of mass be one pound, and if this pound be allowed to fall toward the Earth, at the end of one second it will be moving with the velocity of ($16\frac{1}{12} \times 2$) $32\frac{1}{6}$ feet per second. Now let the mass be a ten-pound weight; it might be thought that, since the Earth attracts each pound of this weight, and therefore attracts the whole with ten times the force with which it attracts one pound, we should have a much greater velocity produced. The old schoolmen thought so; but Galileo showed that a ten-pound weight will fall to the ground with the same velocity as a one-pound weight. This fact is quite consistent with our definitions of gravity and force. Undoubtedly the ten-pound weight is attracted with ten times the force,—but then there is ten times the mass to move; so that, although the *velocity* produced in one second is no greater than in the case of the one-pound

---

what reasoning did he arrive at this conclusion? 502. Is gravity confined to any particular kind of matter? State the law of gravity. By what is the intensity of a force measured? Illustrate this in the case of a one-pound and a ten-pound

weight, yet if we multiply this velocity by the mass the *momentum* produced is ten times as great.

503. Now, since each individual atom of the Earth attracts each individual atom of the weight, we might expect, from our definition of gravity, as well as from the well-known law that every action has a reaction, that the Earth, when the weight is dropped, at the end of one second rises toward the weight with the same momentum that the weight falls to the Earth. No doubt it does; but as the Earth is a very large mass, this momentum represents a velocity infinitesimally small.

504. **Effect of an Increase of Mass in the Attracting Body.**—It follows from the above law that, if the mass of the Earth were twice as great as it now is, it would produce in a falling body twice the present velocity, or $64\frac{1}{3}$ feet per second; and were it only half as great, we should have but half the present velocity produced, or $16\frac{1}{12}$ feet per second.

Accordingly, at the surface of the Moon the force of gravity is very small, whereas on the Sun it is enormous. A man carried to the Moon and retaining the same muscular power, could jump six times as high as on the Earth's surface; whereas, if carried to the Sun, he would be so strongly attracted by its immense mass that he would be literally crushed by his own weight.

505. **Effect of Distance.**—A body at the surface of the Earth, or 4,000 miles from its centre, acquires, as we have seen, by virtue of the Earth's attraction, a velocity of $32\frac{1}{6}$ feet per second at the end of the first second. During this second, however, it has not fallen $32\frac{1}{6}$ feet; for, as it started from a state of rest, and acquired the velocity of

---

weight, and show that the velocity in both cases is the same. 503. What movement might we expect in the Earth, when a weight is dropped? Does the Earth move toward the weight? With what velocity, and why? 504. What is the effect of an increase, and what of a decrease, of mass? What facts are stated with respect to the force of gravity at the surface of the Moon and the Sun? 505. How far does a falling body descend in the first second, near the Earth's surface?

32¼ feet only at the end, it will have gone through the first second with the mean velocity of $16\tfrac{1}{12}$ feet, and will, in fact, have fallen only that distance. Now, this body, at the distance of the Moon, or sixty times as far from the Earth's centre as it now is, would fall in one second toward the Earth only $\tfrac{1}{3600}$ of $16\tfrac{1}{12}$ feet. Let us see how we know this.

506. The Moon's orbit is an exact representation of what the path of our cannon-ball would be at the Moon's

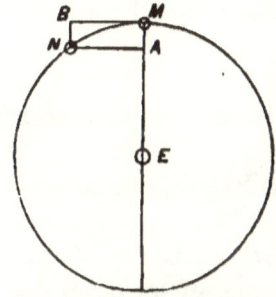

FIG. 104.—ACTION OF GRAVITY ON THE MOON'S PATH.

distance from the Earth. In fact, the Moon's path MN, in Fig. 104, is the result of an *original impulse* in the direction MB, at right angles to EM, and a *constant attraction* toward the Earth—the amount of attraction being represented for the arc MN, by the line MA. To find the value of MA, let us take the arc described by the Moon in one minute, the length of which is found by the following proportion to be nearly 33″:—

27d. 7h. 49m. : 1m. :: 360° : arc described in 1m.

The arc MN, then, being 33″ for one minute of time, the length of MA can be readily calculated; it is found to be $16\tfrac{1}{12}$ feet when ME equals 240,000 miles. That is, a body at the Moon's distance falls as far in one minute as it would do on the Earth's surface in one second; in one second, therefore, by Rule III. Art. 499 (as 60s. make 1m., and 60² = 3600), it will fall but $\tfrac{1}{3600}$ of the distance it would fall in one second at the Earth's surface.

Now, the Moon, being 240,000 miles from the Earth's centre, is just 60 times as far from it as an object at the

Earth's surface is, and we have seen that it is affected by the Earth's attraction only $\frac{1}{3600}$ as much. Its distance is 60 times greater, its gravity is 60^2 times less. Thus the force of attraction is experimentally found *to vary inversely as the square of the distance.* It was this calculation that revealed to Newton the law of universal gravitation.

507. **Kepler's Laws.**—Long before Newton's discovery, Kepler, from observations of the planets merely, had detected certain laws of their motion, which bear his name. They are as follows:—

 I. Each planet describes round the Sun an elliptical orbit, and the centre of the Sun occupies one of the foci.

 II. The radius-vector of a planet describes equal areas in equal times.

 III. If the square of the time of revolution of each planet be divided by the cube of its mean distance from the Sun, the quotient will be the same for all the planets.

508. **Kepler's Second Law.**—It was stated in Art. 301 that the planets move faster as they approach the Sun. Kepler's second law enables us to find how much faster.

The Radius-vector of a planet is the line joining the planet and the Sun. If the planet described a circle, the radius-vector would always be of the same length; but in elliptical orbits its length varies, and the shorter it becomes, the more rapidly does the planet move.

509. In Fig. 105 are shown the orbit of a planet with its eccentricity exaggerated, and the Sun situated in one of the foci. The three shaded areas are equal,—the part of the orbit intercepted being shortest where the radius-vec-

---

lation reveal to Newton? 507. What is meant by Kepler's Laws? Give Kepler's three laws. 508. What was stated in Art. 301? What does Kepler's second law enable us to find? What is the Radius-vector of a planet? In what kind of orbits does the radius-vector vary, and how? 509. Explain the second law, with Fig. 105. Show from the figure how the velocity at perihelion and aphelion must

KEPLER'S LAWS.   279

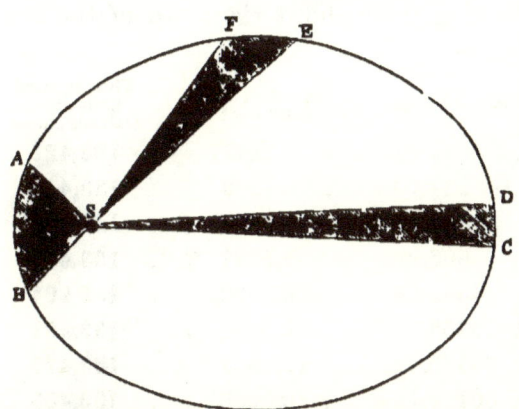

FIG. 105.—ILLUSTRATION OF KEPLER'S SECOND LAW.

tor is longest, as must be the case in order to make the areas equal.

The arcs AB, CD, and EF, are respectively those described at perihelion, at aphelion, and at mean distance, and according to the second law they are traversed in equal times. Therefore, as a greater distance has to be got over at perihelion, and a less one at aphelion, than when the planet is at its mean distance, the motion in the former case must be more rapid, and in the latter case slower, than in other parts of the orbit.

510. **Kepler's Third Law.**—The third law shows that the periodic time of a planet and its distance from the Sun are in some way related; so that, if we represent the Earth's distance and periodic time by 1, and know the period of any planet in terms of the Earth's period, we can at once determine its distance from the Sun in terms of the Earth's distance, by a simple proportion. Thus, in the case of Jupiter:—

$$\left.\begin{array}{c}\text{Square of}\\\text{Earth's}\\\text{period}\\1\times 1\end{array}\right\} : \left\{\begin{array}{c}\text{Square of}\\\text{Jupiter's}\\\text{period}\\11.86\times 11.86\end{array}\right. :: \left.\begin{array}{c}\text{Cube of}\\\text{Earth's}\\\text{distance}\\1\times 1\times 1\end{array}\right\} : \left\{\begin{array}{c}\text{Cube of}\\\text{Jupiter's}\\\text{distance}\\=\\140.559\end{array}\right.$$

That is, whatever the distance of the Earth from the Sun may be, the distance of Jupiter is $\sqrt[3]{140}$ times greater.

compare with that at mean distance. 510. What does Kepler's third law show us? What must we know, to determine a planet's distance from the Sun in terms of the Earth's distance? Illustrate this in the case of Jupiter. 511. What does

511. The following table shows the truth of the law we are considering:—

	Periodic Time.	Mean distance. Earth's = 1.	Time squared divided by distance cubed.
Mercury	87.97	0.3871	133,421
Venus	224.70	0.7233	133,413
Earth	365.25	1.0000	133,408
Mars	686.98	1.5237	133,410
Jupiter	4332.58	5.2028	133,294
Saturn	10759.22	9.5388	133,401
Uranus	30686.82	19.1824	133,422
Neptune	60126.72	30.0368	133,405

512. **Centrifugal and Centripetal Force.**—As these laws were given to the world by Kepler, they simply represented facts, but Newton showed that they all established the truth of the law of gravitation and flowed naturally from it. He proved that the motion of a planet in any part of its orbit is the result of two forces—one *the original impulse*, which gives it a tendency to move off from its orbit in a tangent, and which is called the **Centrifugal Force**—the other *the attraction of the Sun*, which deflects it toward that body, and is called the **Centripetal Force**.

513. Newton also showed that the attraction is proportional to the product of the masses of the bodies. That if we take two bodies, the Sun and our Earth, for instance, we may imagine all the gravitating energies of each to be concentrated at its centre; and that, if the smaller one receives an impulse neither exactly toward nor from the larger, it will describe an orbit round the

---

the table show? How do you find the results to agree? In the case of what planet is there the greatest deviation? 512. What did Newton show with respect to these laws of Kepler? Of what did he prove that the motion of a planet in any part of its orbit is the result? 513. To what did Newton show that the attraction is proportional? Where may we imagine all gravitating energy to be concentrated? What did Newton show with regard to the smaller of two bodies bound together by mutual attraction? What kind of an orbit will it describe? What would follow, if the attraction of the central body were to cease?

CENTRIFUGAL AND CENTRIPETAL FORCE. 281

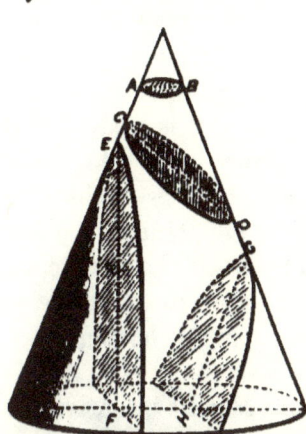

Fig. 106.—The Conic Sections: *AB*, circle; *CD*, ellipse; *EF*, hyperbola; *GH*, parabola.

larger. That this orbit will be one of the conic sections—that is, either a circle, ellipse, hyperbola, or parabola (see Fig. 106). Which of these it will be, depends in each case on the direction and force of the original impulse, which, since the movements of the heavenly bodies are not arrested as bodies in motion on the Earth's surface are, is still at work.

Were the attraction of the central body to cease, the revolving body would leave its orbit, in consequence of the centrifugal tendency it acquired at its start; were the centrifugal tendency to cease, the centripetal force would be uncontrolled, and the body would fall upon the attracting mass.

514. We may now inquire how it is that, according to Kepler's second law, equal areas are traversed in equal times.

The direction of a body moving round another in a circular orbit is always at right angles to the line joining the two bodies. If the orbit be elliptical, the direction is thus perpendicular only at two points; i. e., at the *apsides*, or extremities of the major axis—the aphelion and perihelion points.

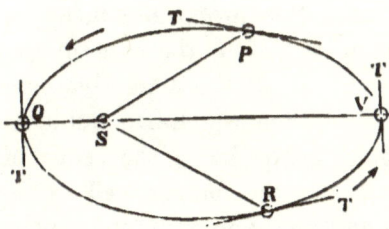

Fig. 107.—Varying Velocity of a Body moving in an Elliptical Orbit, explained.

In Fig. 107, the planet

What, if the centrifugal tendency were to cease? 514. What is always the direction of a body revolving round another in a circle? If the orbit be elliptical, where alone is the direction thus perpendicular? With Fig. 107, explain the

P is moving in the direction PT, the tangent to the ellipse at the place it occupies; this direction not being at right angles to the radius-vector, the attractive force of the Sun helps the planet along. At R it is evident that the attractive force is pulling the planet back. At Q the centripetal force is strong, but the planet is enabled to overcome it by the increased centrifugal tendency it has acquired from being acted on at P. At V the centripetal force is weak, but the planet is not able to overcome it, on account of its centrifugal tendency's having been diminished from being acted on as at R.

515. **Gravity not dependent on the Mass of the Attracted Body.**—We learned in Article 498 that the attraction which a body exerts is the same on all bodies equally distant from it, without reference to their mass. A dime and a feather are equally attracted by the Earth. In like manner, if we had the Sun, Jupiter, a pea, and a mass twice as great as the Sun, at the same distance from the Earth, the Earth's attraction would draw them all through the same number of feet in a second.

516. **Centre of Gravity and Motion.**—Since the amount of attraction is proportioned to the mass of the attracting body (Art. 502), it follows that the attraction of a body with 1 unit of mass will be 1,000 times less than that of a body with 1,000 units of mass—this proportion being, of course, kept up at all distances. If in the case of two bodies, such as the Earth and Sun, all the attractive force were confined, say, to the Sun, then the Earth would revolve round the Sun, the Sun's centre being the centre of motion. But as the Earth draws the Sun, as well as the Sun the Earth, both Earth and Sun revolve round a point in a line joining the two, called the Centre of Gravity.

---

varying velocity of a body moving in an elliptical orbit. 515. Of what is the attraction which a body exerts entirely independent? 516. If all the attractive force were confined to one of two bodies, what would be the result? As they mutually attract each other, what follows? What is meant by the Centre of

517. The centre of gravity and motion would be determined, if we could join the two bodies by a bar, and find the point of the bar at which (supported on a fulcrum) they would balance each other. It is clear that, if the two bodies were of equal mass, this point would be half-way between the two, as at *C* in Fig. 108. If one were heavier than the other, the point of support would approach the heavier body in the ratio of its greater weight (see Fig. 109). In the case of the Sun and Earth, for instance, the centre of gravity of the two lies within the Sun's surface.

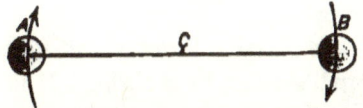
Fig. 108.—Centre of Gravity and Motion in the Case of Equal Masses.

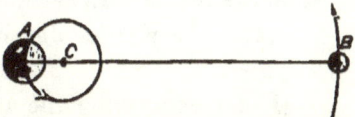
Fig. 109.—Centre of Gravity and Motion in the Case of Unequal Masses.

518. **Determination of Masses.**—It follows, from what has been stated, that the masses of the Sun, and of those planets which have satellites, can be determined, if the mass of our own Earth and the distances of the attracted bodies from their centres of motion are known. Since, for instance, the planets revolve round the Sun, from the curvature of their paths, we can determine the amount of the Sun's attraction,—which, it will be remembered, is proportioned directly to his mass, and is wholly independent of the mass of the attracted body. Having ascertained the Sun's attractive force, and adjusted it to the distance of 4,000 miles from his centre, we can compare it with that of the Earth, and find how many times greater his mass is than the Earth's (Art. 522). In like manner, we can weigh Jupiter, Saturn, Uranus, and Neptune, by finding the effect their attraction has on the orbits of their satellites—also

---

Gravity ? 517. How could the centre of gravity and motion be determined? Where would it lie, if the bodies were of equal mass? Where, if one were heavier than the other? Where does it lie in the case of the Sun and the Earth? 518. How can the masses of the Sun and of those planets that have satellites be determined? How can the masses of those double stars whose distances are known

the double stars whose distances are known, by measuring their effect on each other's orbit.

519. Determination of the Earth's Density and Mass.—We must first find the Earth's mass, or weight. It is not sufficient to determine its bulk, because it might be light, like a gas, or heavy, like lead. The mean density, or *specific gravity*, of its materials—that is, how much they weigh, bulk for bulk, compared with some well-known substance, such as water—must be determined.

520. The Earth's density has been determined in three ways:—

I. By comparing the attractive force of a large metallic ball of known size and density, with that of the Earth.

II. By finding how much a large mountain will deflect a plumb-line, or draw it toward itself from the perpendicular.

III. By determining the rate of vibration of the same pendulum on the top and at the bottom of a mountain, or at the bottom of a mine and at the Earth's surface.

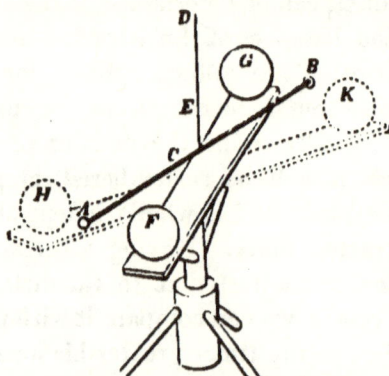

FIG. 110.—THE CAVENDISH EXPERIMENT. *A B*, the small leaden balls on the rod *C*. *D E*, the suspending wire. *F G*, the large leaden balls on one side of the small ones. *H K*, the large leaden balls in a position on the other side.

521. The Cavendish Experiment.—It will here suffice to describe the first-mentioned method, adopted by Cavendish in 1798. The weight of any thing is a measure of the Earth's attraction. Cavendish, therefore, took two small leaden balls of known weight, and fixed them at the ends of a slender wooden rod six feet long, suspended by a fine wire. When the rod was at rest, he placed two large leaden balls one on either side of the small ones. If the large balls exerted any appreciable attractive influence on the smaller ones, the wire would twist to allow each small ball to approach the large one near it; and a telescope was arranged to mark the deviation.

Cavendish found there was a deviation. This enabled him to calculate how great it would have been had each large ball been of the size of the Earth. He then had the attraction of the Earth (measured by the weight of the small balls), and the attraction of a mass of lead as large as the Earth, as the result of his experiment. The density of the Earth, then, was to the density of lead as the attraction of the Earth was to the attractive force of a leaden ball as large as the Earth. This proportion gave the Earth a density of 5.45 as compared with water, the density of lead being 11.35. With this density, the mass of the whole Earth can readily be determined; it amounts in round numbers to

6,000,000,000,000,000,000,000 tons.

But this number is not needed in Astronomy; the relative masses indicated in Art. 157 are sufficient.

522. **Determination of the Sun's Mass.**—We are now prepared to determine the Sun's mass, if we can find how many times it is greater than that of the Earth. This we can do by comparing the action of the Sun and the Earth on a falling body.

On the Earth's surface, i. e., 4,000 miles from its centre, a body falls $16\frac{1}{12}$ feet in a second. Can we determine how far it would fall at 4,000 miles from the centre of the Sun? This is easy: by the process used in the case of the Moon (Art. 506), we find that the Earth falls to the Sun .0099 feet in a second. But this is at a distance of 91,000,000

determined? 521. Give an account of the Cavendish experiment. What was the Earth's density thus found to be? What is its mass? 522. Give the details of

miles from the Sun's centre. We must bring this to 4,000 miles from the Sun's centre, or 22,750 times nearer,—which we do by multiplying the square of 22,750 by .0099, since attraction varies inversely as the square of the distance. The result is 5,123,758 feet. Then

$$16\tfrac{1}{12}\text{ ft.} : 5{,}123{,}758\text{ ft.} :: 1 : \begin{cases}\text{the Sun's mass in}\\ \text{terms of the Earth's.}\end{cases}$$

Solving this proportion, we find that the Sun's mass is approximately 318,641 times greater than that of the Earth. The exact figures are given in Table IV. of the Appendix.

523. Similarly, from the orbit of any of the satellites we determine its rate of fall at 4,000 miles from the centre of its primary, and by the same process as above we find the mass of its primary in terms of the Earth's mass.

524. **Determination of the Comparative Force of Gravity.**—The force of gravity on the surface of the Sun or a planet, compared with that on our Earth, may be determined in the following manner:—

Let us take the case of the Sun. If we express the Sun's mass and radius in terms of the Earth's, then the force of gravity on the Sun's surface, in terms of that on the Earth's surface, will be

$$\frac{\text{Sun's mass}}{\text{Square of radius}} = \frac{314760}{107.8^2} = 27.$$

525. **Perturbations.**—We have seen that it is the attraction of gravitation which causes the planets and satellites to pursue their paths round the central body; that their motion is similar to that of a projectile fired on the Earth's surface, if we leave out of consideration the resistance of the air; and that Newton's law enables us to determine the masses of the Sun and of the other central

---

the process by which the Sun's mass is determined What is it found to be in terms of the Earth's mass? 523. By the same process, what further may we determine? 524. How is the force of gravity on the Sun's surface found? 525. What have we found with respect to the motion of the planets and satellites?

bodies from the motions of the bodies revolving round them, when the mass of the Earth itself is known.

Now, the orbit which each body would describe round the Sun or round its primary, if itself and the Sun or primary were the only bodies in the system, is liable to variations in consequence of the attractions of the other planets and satellites. These irregular attractions, which vary according to the constantly-changing distances between the bodies, are called Perturbations, and the resulting changes in the motions of the bodies affected are called Inequalities if the disturbances are large, and Secular Inequalities if they extend over a long period of time.

526. These perturbations and inequalities are among the most difficult subjects in the whole domain of astronomy; it is sufficient here to say that it is by carefully observing them that we have been able to determine the masses of the planets that have no satellites and of the satellites themselves.

527. We shall conclude with an explanation of two additional and very important effects of attraction of a somewhat different kind. One results from the attractions of the Sun and Moon on the equatorial protuberance of the Earth, and is called the Precession of the Equinoxes; the other is due to the attraction of the water on the Earth's surface by the Sun and Moon, whence result the Tides.

528. **Precession of the Equinoxes, how produced.**—Let the equatorial protuberance of the Earth be represented by a ring, supported by two points at the extremities of a diameter, and inclined to its support as the Earth's equator is inclined to the ecliptic. Let a long string be attached to the highest portion of the ring, and be pulled horizontally, at right angles to the line connecting the

---

What does Newton's law enable us to do? What is meant by Perturbations? What are Inequalities? What are Secular Inequalities? 526. By carefully observing these inequalities, what have we been enabled to determine? 527. What two effects of attraction remain to be considered? From what does each result? 528. Illustrate the effect of the Sun's attraction in producing precession, with a

288 UNIVERSAL GRAVITATION.

two points of suspension, and away from the centre of the ring. This pull will represent the Sun's attraction on the protuberance. The effect on the ring will be that it will at once take a horizontal position; the highest part of the ring will fall as if it were pulled from below, the lowest part will rise as if pulled from above.

The Sun's attraction on the equatorial protuberance in certain parts of the orbit is exactly similar to the action of the string on the ring, but the problem is complicated by the two motions of the Earth. In the first place, in consequence of the yearly motion, the protuberance is presented to the Sun differently at different times, so that twice a year (at the solstices) his action is greatest, and twice a year (at the equinoxes) it is reduced to 0. In the second place, the Earth's rotation is constantly varying that part of the equator subjected to the attraction.

529. If the Earth were at rest, the equatorial protuberance would soon settle down into the plane of the ecliptic; in consequence, however, of its two motions, this result is prevented, and the attraction of the Sun on a particle situated in the protuberance is limited to causing that particle to meet the plane of the ecliptic earlier than it otherwise would do. If we look at the Earth as presented to the Sun at the winter solstice (Fig. 44), and bear in mind that the Earth's rotation is from left to right in the diagram, it will be clear, that, while the particle is mounting the equator, the Sun's attraction is pulling it down; so that the path of the particle is really less steep than the equator is represented in the diagram. Toward the east, the particle descends from this less height more rapidly than it would otherwise do, as the Sun's attraction

---

ring and a string. By what is the problem complicated? What is the consequence of the yearly motion? What is the consequence of the Earth's rotation? 529. If the Earth were at rest, what would the equatorial protuberance soon do? What is the effect of the Earth's two motions? If we look at the Earth as presented to the Sun at the winter solstice in Fig. 44, what will appear? What does

is still exercised. The final result therefore is, that it meets the plane of the ecliptic sooner than it would otherwise have done.

What happens with one particle in the protuberance happens with all. One half of it, therefore, tends to fall, the other half to rise; and the whole Earth meets the strain by rolling on its axis. The inclination of the protuberance to the plane of the ecliptic is not altered; but, in consequence of the rolling motion, the places in which it crosses that plane precede those at which the equator would cross it were the Earth a perfect sphere; hence the term *precession*.

530. In the above explanation, no mention has been made of the sphere enclosed in the equatorial protuberance, as the action of the Sun on the spherical portion is constant. It plays an important part, however, in averaging the precessional motion of the entire planet during the year, acting as a brake at the solstices, when the Sun's effect on the equatorial protuberance is greatest, and continuing the motion at the equinoxes, when, as before stated, the Sun's action is reduced to 0.

We have also, for the sake of greater clearness, left the Moon out of view, although our satellite plays the greatest part in precession, for the following reason. The action referred to does not depend on the actual attractions of the Sun and Moon upon the Earth as a whole, which are in the proportion of 190 to 1, but on the *different degrees of attraction exerted by each upon different parts of the Earth*. As the Sun's distance is so great compared with the diameter of the Earth, the differential effect of the Sun's action is small; but, as the Moon is so near, her differential effect and consequent influence in producing precession is three times that of the Sun.

---

one-half of the protuberance tend to do, and what the other? How does the Earth meet the strain? What is the consequence of the rolling motion? 530. What is the effect of the sphere enclosed in the equatorial protuberance? What part does

531. Change in the Earth's Axis.—The change in the position of the equator which follows from the rolling motion, is necessarily connected with a change in the Earth's axis. This change consists in a slow revolution round the axis of the celestial sphere, perpendicular to the plane of the ecliptic.

532. Nutation.—Superadded to the general effect of the Sun and Moon in causing the precession of the equinoxes, is an effect due to the Moon alone, termed Nutation.

The Moon's nodes perform a complete revolution in nineteen years. Consequently, for half this period the Moon's orbit is less inclined to the plane of the Earth's equator than the ecliptic is; during the other half, the orbit is inclined so that its divergence from the plane of the Earth's equator is the greatest possible. In the former position the precessional effect will be small, while in the latter it will be the greatest possible.

Were the pole of the earth at rest, nutation would cause it to describe a small ellipse every nineteen years. Since, however, the pole is in motion, as we just saw in

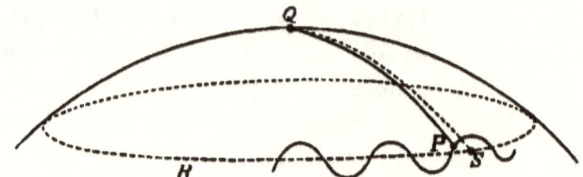

Fig. 111.—Path of the Pole of the Equator, *P*, round the Pole of the Heavens (on Ecliptic), *Q*.

Art. 531, the two motions are compounded, so that the path of the pole of the equator round the pole of the eclip-

---

the Moon perform in producing precession? Why is its effect greater than that of the Sun? 531. What motion is produced in the Earth's axis, in consequence of the change in the position of the equator? 532. What effect, due to the Moon alone, helps to produce the precession of the equinoxes? What is meant by Nutation? If the pole of the Earth were at rest, what would nutation cause it

tic, instead of being circular, is *waved*, as shown in Fig. 111.

533. The effect of this motion of the Earth's axis on the apparent position of the heavenly bodies, and the corrections which are thereby rendered necessary, have already been referred to in Art. 459.

534. **Tides.**—Tides are alternate risings and fallings of the waters of the ocean.

The waters gradually rise for about six hours, forming *flood tide*,—remain stationary a few moments at *high tide*, —then begin to fall, forming *ebb tide*,—reaching *low tide* in about six hours; and then, after a few minutes' rest, these movements are repeated. The interval between two successive high or low tides is 12 hours 27 minutes,—that is, they rise and fall twice in a lunar day.

535. **Spring and Neap Tides.**—We not only have two tides in a lunar day, but twice in the lunar month—from one and a half to two days after new and full Moon—the tides are higher than usual: these are the Spring Tides. Twice also, after the Moon is in her quadratures, they are lower than usual: these are the Neap Tides. It will be gathered from the foregoing that the tides have something to do with the Moon. In fact, these phenomena are due to the attraction of the Sun and Moon on the fluid envelope of the Earth—not to their absolute, but (as in the case of precession) to their differential, action; and the two periods correspond with the *lunar* day and the *lunar* month, because the Moon's differential attraction is about three times as great as that of the Sun.

536. **Tides, how produced.**—It may be stated, then, generally, that the semi-diurnal tides are caused by the Moon (although there is really a smaller daily tide caused

to do? Since it is in motion from another cause, what is the consequence? 534. What are Tides? Describe the succession of tides. What is the interval between two successive high or low tides? Why is it just 12 hours and 27 minutes? 535. What is meant by Spring Tides and Neap Tides? To what are tides due? Why are they specially connected with the lunar day and month? 536. How are

by the Sun); and that the semi-monthly variation in their height is due to the Sun's tide being added to that of the Moon when she is new and full, at which time the Sun and Moon pull together,—and subtracted from it at the first and last quarters, when they are pulling at right angles to each other.

The spring and neap tides thus produced are also affected by the difference of latitude between the two bodies. Of course, that spring tide will be highest which occurs when the Moon is nearest her node, or in the ecliptic. The apex of the semi-diurnal tide, also, follows the Moon throughout her various declinations.

537. The double daily tide arises from the action of the Moon on both the water and the Earth itself. On the side toward the Moon, the water is pulled from the Earth and piled up under the Moon, as the Moon's action on the surface-water is greater than its action on the Earth's centre, which is more remote. In like manner, since the Moon's attraction for the Earth's centre is greater than its attraction for the water on the opposite side of the Earth, and since the solid Earth must move with its centre, the Earth is pulled from the water. Hence there are always two tides on the Earth's surface; and this double tide is simply a state of the water, without progressive motion and nearly at rest under the Moon. There is, in fact, an ellipsoid of water enclosing the Earth, which always remains with its longer axis pointing to the Moon.

538. The high water under, or nearly under, the Moon, is not caused merely by the direct attraction of our satellite acting upon the particles immediately beneath it, but by its action on all the particles of water on the side of the Earth turned to it, all of which tend to *close up* under

---

ordinary tides, and the spring and neap tides, produced? By what are the spring and neap tides also affected? 537. Explain why there is a double daily tide. 538. What besides the direct attraction of the Moon on the particles immediately beneath it helps to produce the tides? What is meant by the tangential component

the Moon. The force acting upon these particles is called the *tangential component* of the attraction; and this is by far the most powerful cause of the tides, as it acts at right angles to the Earth's gravity, whereas the direct attraction of the Moon acts in opposition to this gravity.

539. **Establishment of the Port.**—The phenomena of the tides are greatly complicated by the irregular distribution of land. The time of high water at any one place occurs at the same period from the Moon's passage over the meridian; this period is different for different places. The interval at new or full Moon between the time of the Moon's meridian passage and high water is termed the Establishment of the Port.

540. **Velocity and Height of the Tidal Wave.**—In the open ocean, the velocity of the tidal wave may be as great as 900 miles an hour; in shallow waters, it may be retarded to even 7 miles, while its height may be greatly increased. The average height of the tide round the islands in the Atlantic and Pacific Oceans is but $3\frac{1}{2}$ feet; whereas at the head of the Bay of Fundy it is 70 feet.

541. **Effect of Tidal Action on the Daily Rotation.**—As the tidal wave, being regulated by the Moon, does not move so rapidly as the Earth, it appears to move westward while the Earth is moving eastward; and it has been suggested that this movement *acts as a brake* on the Earth's daily rotation, causing a constant but very slight decrease in its velocity. The apparent acceleration of the Moon's mean motion may be accounted for by supposing that the sidereal day is shortening, in consequence of tidal action, at the rate of $\frac{1}{60}$ of a second in 2,500 years.

of the attraction? 539. By what are the phenomena of the tides greatly complicated? What is meant by the Establishment of the Port? 540. How great may the velocity of the tidal wave be in the open ocean? What is it sometimes in shallow waters? What is the average height of the tide round the islands in the Atlantic and Pacific? At the head of the Bay of Fundy? 541. What is the effect of tidal action on the Earth's daily rotation? How much may we suppose the sidereal day to be shortened in consequence of tidal action?

APPENDIX.

Table I.
EXPLANATION OF ASTRONOMICAL SYMBOLS.

Signs of the Zodiac.

		°			°
0. ♈ Aries		0	VI. ♎ Libra		180
I. ♉ Taurus		30	VII. ♏ Scorpio		210
II. ♊ Gemini		60	VIII. ♐ Sagittarius		240
III. ♋ Cancer		90	IX. ♑ Capricornus		270
IV. ♌ Leo		120	X. ♒ Aquarius		300
V. ♍ Virgo		150	XI. ♓ Pisces		330

The Sun . . . . ☉   A Comet . . . . ☄
The Moon . . . . ☽   A Star . . . . . ✴

☌ Conjunction.
☐ Quadrature.
☍ Opposition.
☊ Ascending Node.
☋ Descending Node.
h. Hours.
m. Minutes of Time.
s. Seconds of Time.

° Degrees.
′ Minutes of Arc.
″ Seconds of Arc.
R.A., or ℞., or α., Right Ascension.
Dec^l., D., or δ, Declination.
N. P. D., North-polar Distance.

Greek Alphabet, used in naming the Stars.

α	Alpha.	ι	Iota.	ρ	Rho.
β	Beta.	κ	Kappa.	σ	Sigma.
γ	Gamma.	λ	Lambda.	τ	Tau.
δ	Delta.	μ	Mu.	υ	Upsilon.
ε	Epsilon.	ν	Nu.	φ	Phi.
ζ	Zeta.	ξ	Xi.	χ	Chi.
η	Eta.	ο	Omicron.	ψ	Psi.
θ	Theta.	π	Pi.	ω	Omega.

Major Planets.

☿ Mercury.
♀ Venus.
⊕ or ♁ The Earth.
♂ Mars.

♃ Jupiter.
♄ Saturn.
♅ Uranus.
♆ Neptune.

ASTEROIDS, OR MINOR PLANETS.

1. Ceres.
2. Pallas.
3. Juno.
4. Vesta.
5. Astræa.
6. Hebe.
7. Iris.
8. Flora.
9. Metis.
10. Hygeia.
11. Parthenope.
12. Victoria.
13. Egeria.
14. Irene.
15. Eunomia.
16. Psyche.
17. Thetis.
18. Melpomene.
19. Fortuna.
20. Massilia.
21. Lutetia.
22. Calliope.
23. Thalia.
24. Themis.
25. Phocea.
26. Proserpine.
27. Euterpe.
28. Bellona.
29. Amphitrite.
30. Urania.
31. Euphrosyne.
32. Pomona.
33. Polyhymnia.
34. Circe.
35. Leucothea.
36. Atalanta.
37. Fides.
38. Leda.
39. Lætitia.
40. Harmonia.
41. Daphne.
42. Isis.
43. Ariadne.
44. Nysa.
45. Eugenia.
46. Hestia.
47. Aglaia.
48. Doris.
49. Pales.
50. Virginia.
51. Nemausa.
52. Europa.
53. Calypso.
54. Alexandra.
55. Pandora.
56. Melete.
57. Mnemosyne.
58. Concordia.
59. Olympia.
60. Echo.
61. Danaë.
62. Erato.
63. Ausonia.
64. Angelina.
65. Maximiliana.
66. Maia.
67. Asia.
68. Leto.
69. Hesperia.
70. Panopea.
71. Niobe.
72. Feronia.
73. Clytie.
74. Galatea.
75. Eurydice.
76. Freia.
77. Frigga.
78. Diana.
79. Eurynome.
80. Sappho.
81. Terpsichore.
82. Alcmene.
83. Beatrix.
84. Clio.
85. Io.
86. Semele.
87. Sylvia.
88. Thisbe.
89. Julia.
90. Antiope.
91. Ægina.
92. Undina.
93. Minerva.
94. Aurora.
95. Arethusa.
96. Ægle.
97. Clotho.
98. Ianthe.
99. Dike.
100. Hecate.
101. Helena.
102. Miriam.
103. Hera.
104. Clymene.
105. Artemis.
106. Dione.
107. Camilla.
108. Hecuba.
109. Felicitas.
110. Lydia.
111. Ate; etc.

TABLE II.—ELEMENTS OF THE PLANETS.

Symbol.	Distance from the Sun.			Inclination of the Planet's Equator to its Orbit.	Ascending Node of Equator on Orbit.	Equatorial Diameter.	Planet's Mass, Earth's = 1.	Bodies fall in 1 Second	Density, Earth's = 1.	Volume, Earth's = 1.	Distance from the Earth.		Time of Revolution round the Sun.	Apparent Diameter as seen from the Earth.		Synodic Revolution.
	Mean.	Greatest.	Least.								Greatest.	Least.		Greatest.	Least.	
	Miles.	Miles.	Miles.	° ′	° ′	Miles.		Feet.			Miles.	Miles.	Mean Solar Days.			Mean Solar Days.
☿	35,393,000	42,666,000	28,120,000	49 58	56 30	2,962	0.065	7.45	1.24	0.052	135,681,000	47,229,000	87.9692	11.5	4.5	115.887
♀	66,131,000	66,566,000	65,677,000	28 27 24	0 0	7,510	0.785	14.06	0.92	0.851	159,551,000	23,309,000	224.7007	62.0	9.5	583.920
⊕	91,430,000	92,965,000	89,895,000	28 51 0	79 1	7,926	1.000	16.08	1.00	1.000	....	....	365.2568	...	...	...
♂	189,312,000	152,284,000	126,341,000	3 4 0	818 22	4,920	0.124	4.88	0.96	0.189	245,249,000	62,889,000	686.9794	23.5	3.8	779.936
♃	475,693,000	496,604,000	452,783,000	26 49 0	171 43	85,390	300.857	88.89	0.22	1887.431	591,569,000	408,709,000	4332.5848	46.0	30.0	398.867
♄	872,185,000	921,105,000	823,164,000	100 20 0	165 15	71,904	90.082	17.59	0.12	746.898	1,014,071,000	881,210,000	10759.2197	20.5	14.6	378.090
♅	1,758,851,000	1,885,701,000	1,672,001,000	?	?	33,024	12.441	11.71	0.18	72.859	1,928,666,000	1,745,806,000	30686.8205	4.3	3.5	369.656
♆	2,746,271,000	2,770,217,000	2,722,325,000	?	?	36,620	16.761	12.62	0.17	98.664	2,863,183,000	2,629,360,000	60126.722	2.7	2.6	367.488

Symbol.	Time of Rotation on Axis.
	h. m. s.
☿	24 5 28
♀	23 16 19
⊕	23 56 4
♂	24 37 23
♃	9 55 28
♄	10 29 17
♅	?
♆	?

APPENDIX. 297

TABLE III.—ELEMENTS OF THE SATELLITES.

Primary.	No.	Name of Satellite.	Distance from Primary.	Sidereal Revolution.			Inclination of Orbit to Plane of Ecliptic.			Diameter.	Apparent Star Magnitude.	Name of Discoverer.
				d.	h.	m.	°	'	"			
Earth.	1	Moon.	*Miles.* 237,640	27	7	43	5	8	40	*Miles.* 2,158		
Mars.	1	....	6,000	0	7	88				30 ?	12	Prof. A. Hall.
	2	....		0	30	14				15 ?	?	"
Jupiter.	1	Io.	267,360	1	18	27	3	4	6	2,252	7	Galileo.
	2	Europa.	425,160	3	13	14	3	5	5	2,099	7	
	3	Ganymede.	678,360	7	3	43	3	9	2	3,436	6	
	4	Callisto.	1,192,820	16	16	32	3	23	0	3,057	7	
Saturn.	1	Mimas.	120,800	0	22	37				1,000	17	Sir W. Herschel.
	2	Enceladus.	155,000	1	8	53				?	15	"
	3	Tethys.	191,000	1	21	18				500	13	Cassini.
	4	Dione.	246,000	2	17	41				500	12	"
	5	Rhea.	343,000	4	12	25				1,200	10	"
	6	Titan.	796,040	15	22	41	23	10	22	3,300	8	Huyghens.
	7	Hyperion.	1,007,000	21	7	8				?	17	Bond and Lassell.
	8	Japetus.	2,314,000	79	7	55				1,500	9	Cassini.
Uranus [Motion of Satellites retrograde].	1	Ariel.	123,000	2	12	28	100° 34' Ascending Node			?	?	Lassell.
	2	Umbriel.	171,000	4	3	27				?	?	O. Struve.
	3	Titania.	281,400	8	16	55	165° 30'			?	?	Sir W. Herschel.
	4	Oberon.	376,000	13	11	6				?	?	"
Neptune [Motion retrograde].	1	....	220,000	5	21	8	150°			?	14	Lassell.

Table IV.—THE SUN.

	Old Value.	New Value.
Equatorial Horizontal Parallax	8.5776″	8.940″
Mean Distance from the Earth	95,274,000	91,430,000
Diameter in miles	888,646	852,584
Inclination of Axis to Plane of Ecliptic	82° 45′	for 1850
Longitude of Node	73 40	
Mass	354,936	314,760
Density (Earth's taken as 1)	0.250	0.250
Volume (Earth's taken as 1)	1,415,225	1,245,126
Force of Gravity at Equator	28.7	27.2

Time of Rotation . . . . . . Variable with the Latitude.
Apparent Diameter as seen from the Earth:—
    Maximum . . . . 32′ 36.41″
    Minimum . . . . 31′ 32″
    Mean . . . . 32′ 4.205″

Table V.

ADDITIONAL ELEMENTS OF THE MOON.

Mean Horizontal Parallax	57′ 2.70″
Mean Angular Telescopic Semi-diameter	15 34.4
Ascending Node of Orbit	13° 53′ 17″
Mean Synodic Period	29.530588715^d
Time of Rotation	27.321661418^d
Inclination of Equator to Plane of Ecliptic	1° 30′ 10.8″
Longitude of Pole	?
Daily Geocentric Motion	13° 10′ 35″
Mean Revolution of Nodes	6793.39108^d
Mean Revolution of Apogee or Apsides	3232.57343^d
Density, Earth's as 1	0.56654
Volume, "	0.02012
Force of Gravity at surface, Earth's as 1	¼
Bodies fall in one second	2.6 feet.

APPENDIX.

Table VI.—TIME.

I.—THE YEAR.

	Mean Solar Days.			
	d.	h.	m.	s.
The Mean Sidereal Year	365	6	9	9.6
The Mean Solar or Tropical Year	365	5	48	46.054440
The Mean Anomalistic Year	365	6	13	49.3

II.—THE MONTH.

Lunar or Synodic Month	29	12	44	2.84
Tropical Month	27	7	43	4.71
Sidereal "	27	7	43	11.54
Anomalistic "	27	13	18	37.40
Nodical "	27	5	5	35.60

III.—THE DAY.

The Apparent Solar Day, or interval between two transits of the Sun over the meridian . *variable.*
The Mean Solar Day, or interval between two transits of the Mean Sun over the meridian 24 0 0
The Sidereal Day 23 56 4.09
The Mean Lunar Day 24 54 0

Table VII.—CORRECTION FOR REFRACTION.

Apparent Altitude.	Mean Refraction.	Apparent Altitude.	Mean Refraction.	Apparent Altitude.	Mean Refraction.
° '	' "	° '	' "	° '	' "
0 0	34 54	7 0	7 20	25 0	2 3
0 20	30 52	8 0	6 30	30 0	1 40
0 40	27 23	9 0	5 49	35 0	1 22
1 0	24 25	10 0	5 16	40 0	1 9
2 0	18 9	11 . 0	4 49	45 0	0 58
3 0	14 15	12 0	4 25	50 0	0 48
3 30	12 48	13 0	4 5	60 0	0 33
4 0	11 39	14 0	3 47	70 0	0 21
5 0	9 47	15 0	3 32	80 0	0 10
6 0	8 23	20 0	2 37	90 0	0 0

ALPHABETICAL INDEX

AND

ETYMOLOGICAL VOCABULARY OF ASTRONOMICAL TERMS.

[The numbers refer to Articles, not to Pages]

Abbreviations used in Astronomy, see Appendix, Table I.
Aberration of light (*ab*, from, and *errare*, to wander), 410; results of, 452; correction for, 452; constant of, 452; spherical and chromatic, of lenses, 425, 426.
Absorption of the atmospheres of stars, 84; of Sun, 124.
Acceleration, Secular, of the Moon's mean motion, an increase in its velocity caused by a slow change in the eccentricity of the Earth's orbit, 541.
Achromatism (α, without, and χρῶμα, color) of lenses, 424.
Adams, discovered Neptune, 111.
Aerolites (ἀήρ, the air, and λίθος, a stone), meteoric stones which fall to the Earth's surface, 317.
Aerosiderites (ἀήρ and σίδηρος, iron), pieces of meteoric iron which fall to the Earth's surface, 317.
Aerosiderolites, (ἀήρ, σίδηρος, and λίθος), pieces of meteoric iron and stone which fall to the Earth's surface, 317.
Aigrettes, rays of light seen through the corona in a total eclipse of the Sun, 219.
Air, refraction of the, 411.
Algol, the variable star, 75.
Almanac, Nautical, 463.
Alphabet, Greek, 52.

Altazimuth (contraction of *altitude* and *azimuth*), an instrument for measuring altitudes and azimuths, 442; when used, 441.
Altitude (*altitudo*, height), the angular height of a celestial body above the horizon, 332.
Anaximander, conceived the idea of a plurality of worlds, 21.
Angle (*angulus*, a corner), the difference in direction of two straight lines that meet, 26; how named, 26; right, obtuse, and acute, defined, 27; measurement of angles, 438; of position (439), the angle formed by the line joining the components of double stars, with the direction of the diurnal motion. It is reckoned in degrees from the north point passing through east, south, and west.
Angle of the vertical, the difference between astronomical and geodetical latitude. It is 0 at the equator and at the poles, and attains a maximum of 11′ 30″ in lat. 45°.
Annular (*annulus*, a ring), eclipses, 244; nebulæ, 95.
Anomalistic, month, 399, 400; year, 401, 402.
Anomaly (α, not, and ὁμαλός, equal). The anomaly is either true, mean, or eccentric. The first is the true distance of a planet or comet from peri-

ALPHABETICAL INDEX. 301

helion; the second, what it would have been had it moved with a mean velocity; and the third, an auxiliary angle introduced to facilitate the computation of a planet's or comet's motion.

Ansæ (Lat. *handles*) of Saturn's rings, 282.

Anti-trades, 207-209.

Aphelion (ἀπό, from, and ἥλιος, the Sun), the point in an orbit farthest from the Sun, 175; distance of comets, 298; of planets, see Appendix, Table II.; change of, 407.

Apogee (ἀπό and γῆ, the Earth). (1) The point in the Moon's orbit farthest from the Earth, 218. (2) The position in which the Sun or other body is farthest from the Earth.

Apsis (ἁψίς, a curve), plural **Apsides**. The line of apsides (407) is the line joining the aphelion and perihelion points; it is therefore the major axis of elliptic orbits.

Arabians, the chief astronomers of the Dark Ages, 22.

Aratus, described the leading constellations in verse, 54.

Arc, Diurnal, the path described by a heavenly body between rising and setting, 360; Semi-diurnal, half this path on either side of the meridian.

Arcturus, proper motion of, 63.

Aries, First Point of, one of the points of intersection of the celestial equator and ecliptic, and the starting-point for R. A. and celestial longitude, 328.

Aristarchus, taught that the planets revolve round the Sun, 21.

Aristotle, his opinion respecting the Milky Way, 46.

Ascending Node, 221.

Ascension, Right, the angular distance of a heavenly body from the first point of Aries, measured on the equator, 326.

Aspects of the planets, 370.

Asteroids (ἀστήρ, a star, and εἶδος, form), minor planets, 137; discovery of, 142, 291; size, 292; force of gravity, 292; orbits, 293; evidences of atmosphere and rotation, 294; mode of discovery, 295; theory respecting, 296.

Astronomy (ἀστήρ, a star, and νόμος,

a law), defined, 1; usefulness of, 14; early history of, 19-23.

Atmosphere (ἀτμός, vapor, and σφαῖρα, a sphere), of stars, 82; of Sun, 124, 126; of Earth, 205-211, 214; of Moon, 233; of Mars, 270; of Jupiter, 277; refraction of the, 411.

Attraction of gravitation, *see* Gravitation.

Axis, the line on which a heavenly body rotates: the major axis of an elliptical orbit is the line of apsides; the minor axis is the line at right angles to it; the semi-axis major is equal to the mean distance.

Axis of the Earth, its inclination, 180; motion of, 531; inclination of Sun's, 112.

Axis, polar, and declination, of equatorials, 436.

Azimuth (*samatha*, Arabic, to go toward), the angular distance of a celestial object from the north or south point of the meridian, 332.

Baily's Beads, notched appearance sometimes presented in the narrow ring of light in an annular eclipse, 250.

Base-line, 469.

Bayer, John, his method of naming the stars, 58.

Belts, of Jupiter, 276; of calms and winds, 207; of Saturn, 281.

Biela's Comet, 303.

Bissextile (*bis*, twice, and *sextus*, sixth), 403.

Bode's Law, 290.

Bolides, 307.

Bond, his discoveries respecting Saturn's rings, 282.

Brilliancy, of the stars, 41; Sun, 106; Moon, 234; minor planets, 295.

Calendar, 403-405.

Calms, equatorial, 207; of Cancer and Capricorn, 207; polar, 207.

Catalogues of stars, 440.

Cavendish experiment, the, 521.

Celestial, sphere, 326-329; apparent movements of, 334; two methods of dividing, 357; meridian, 333.

Centre (κέντρον) of gravity, 516, 517.

Centrifugal Force, 512, 513.

ALPHABETICAL INDEX.

Centripetal Force, 512, 513.
Chaldeans, their early progress in Astronomy, 20.
Chinese, their early attention to Astronomy, 20.
Chronograph (χρόνος, time, and γράφω, I write), an instrument for determining the time of transit of a heavenly body across the field of view of a transit-circle or other instrument, 447.
Circle, defined, 31; great and small, defined, 38; declination, the circle on the declination axis of an equatorial, by which the declinations of celestial bodies are measured, 436; transit, an instrument for observing the transit of heavenly bodies across the meridian and their zenith-distance, 443; of perpetual apparition, a circle of polar distance equal to the latitude of the place, the stars within which never set, 339.
Circumference of a circle, defined, 31; how divided, 32.
Circumpolar stars, 338.
Clepsydræ (κλεψύδρα, a water-clock), 384.
Clock, principles of construction, 388; invention of, 389; Tycho Brahe's regulator, 389; application of the pendulum, 389; sidereal, 397.
Clock-stars, stars the positions of which have been accurately determined, used in regulating astronomical clocks and determining the time.
Clouds, on the Earth, 211; on Mars, 270.
Clusters of stars, 87, 88.
Co-latitude of a place or a star is the difference between its latitude and 90°.
Collimation (*cum*, with, and *limes*, a limit), line of the optical axis of a telescope; error of, the distance of the cross-wires of a telescope from the line of collimation, 438.
Collimator, a telescope used for determining the line of collimation in fixed astronomical instruments.
Colors of stars, 78–80.
Colures (κολούω, I divide), meridians passing through the equinoxes and solstitial points, called the equinoctial and solstitial colures.
Coma (Lat. *hair*) of a comet, 300.

Comes (Lat. *companion*), the smaller component of a double star.
Comets (κομήτης, long-haired), probably masses of gas, 13, 304; orbits of, 298; distances from Sun, 298, 299; long and short period comets, 298; head, nucleus, and tail, 300; changes as they approach the Sun, 301; envelopes, 301; velocity of, 301; probably harmless, 302; division of Biela's comet, 303; physical constitution of, 304; numbers of, in our system, 305; how formerly regarded, 306.
Compression, polar, the amount by which the polar diameter of a planet is less than its equatorial one; of the Earth, 202, 203; of Mercury, 260; of Venus, 264; of Jupiter, 273.
Cone of shadow in eclipses, 241.
Conic sections, the, 513.
Conjunction. Two or more bodies are said to be in conjunction when they are in the same longitude or right ascension. In inferior conjunction, the bodies are on the same side of the Sun; in superior conjunction, on opposite sides, 370.
Constant of aberration, 452.
Constellation (*cum*, with, and *stella*, a star), a group of stars supposed to represent some figure, 53; by whom arranged, 54; zodiacal, 55; northern, 56; southern, 57; circumpolar, 342, 343; visible on different evenings throughout the year, 349–351.
Copernicus, discovered the true system of the universe, 22.
Copernicus, lunar crater, 230.
Corona (Lat. *crown*), the halo of light which surrounds the dark body of the Moon during a total eclipse of the Sun, 249.
Corrections applied to observed places, 450–460; for refraction, 451; aberration, 452; parallax, 454; precession of the equinoxes and nutation, 456–460.
Corrugations, on the Sun's disk, 120.
Cosmical rising and setting of a heavenly body, its rising or setting with the Sun.
Craters (κρατήρ, a mixing-bowl) of the Moon, 227–230.
Crust of the Earth, 193–197; tempera-

ALPHABETICAL INDEX. 303

ture of, 199; thickness, 200; density, 201.

Culmination (*culmen*, the top), the passage of a heavenly body across the meridian, when it is at the highest point of its diurnal path.

Curtate distance, the distance of a celestial body from the Sun or Earth projected on the plane of the ecliptic.

Cusp (*cuspis*, a sharp point), the extremities of the illuminated side of the Moon or inferior planets at the crescent phase.

Cycle (κύκλος, a circle) of eclipses, 247.

Dawes, his description of the red-flames seen during the total eclipse of the Sun in 1860, 251; discovery of Saturn's inner ring, 282.

Day, apparent and mean solar, 395; civil, 395; sidereal and solar, 353; and night, how produced, 182, 183; length of, in different latitudes, 184; how to find the length of. 361.

Declination, the angular distance of a celestial body north or south from the equator, 328; parallels of, 328; axis of equatorials, 436.

Degree, the 360th part of any circle, 32.

De La Rue, Mr., his lunar, solar, and planetary photographs, 495.

Democritus, his opinion respecting the Milky Way, 46.

Density, what it is, 155; of the Earth, 156; how determined, 520; of the Earth's crust, 201; of the Sun, 109; of the planets, 157.

Descending Node, 221.

Detonating meteors, 316.

Diameter, of the Earth, 162; Moon, 217; Sun, 108; planets, 150; of heavenly bodies, how found, 475.

Digit, the twelfth part of the diameter of the Sun or Moon, used in measuring the extent of a partial eclipse, 246.

Dimensions, of the Sun, 108; Earth, 162; Moon, 217; lunar craters, 227; the planets, 150; Saturn and his rings, 284; how determined, 475.

Diodorus, his opinion respecting the Milky Way, 46.

Disk (δίσκος), the visible surface of the Sun, Moon, or planets, 105.

Dispersion of light, 414; varies in different substances, 415.

Distance, of stars, 43, 44; how determined, 474; of nebulæ, 29; of Sun, 107; how determined, 472; old and new value of, 472; of planets from Sun, 148; how determined, 469; of asteroids, 293; of Moon from Earth, 218; polar, 329; how distances are measured, 469.

Donati's Comet, 300.

Double Stars, 66, 68.

Earth, the, a planet, 11; is round, 160; rotation proved by Foucault, 163, 164; proved by the gyroscope, 165, 166; poles, 162; equator, 162; diameters, 162; parallels and meridians, 167; latitude, 168; longitude, 169; tropics, polar circles, and zones, 170; length of polar and equatorial diameter, 171; shape, 171, 172, 202, 203; motions of, 173; effects of motions, 181; shape of orbit, 174, 176; change in, 406; eccentricity of orbit, 176; when in perihelion, 176; velocity of rotation, 177; velocity of revolution, 178; inclination of axis, 180; day and night, 182; how caused, 183; length of. 184; how to determine, 361; seasons, 186-189; structure and past history, 192-197; crust, of what composed, 193; interior temperature of, 198, 199; once a star, 197; density of the crust, 201; atmosphere, 205-211; belts of calms and winds, 207; cause of winds, 210; elements in the Earth's crust, 212, 213; in the Earth's atmosphere, 214; appearance, as seen from Moon, 218. *Apparent movements.*—The Earth, the centre of the visible creation, 324; apparent movements of the heavens, due to the real movements of the, 325; effects of rotation, 325, 344; apparent movements of the stars, as seen from different points on the surface, 335-337; effects of the Earth's yearly motion, 345, 346; Earth's way, 453; effects of attraction of, 501; motion of axis, 531.

Earth-shine, 223.

Eccentricity (*ex*, from, and *centrum*, a centre), of an ellipse, the distance of either focus from the centre, divided by half the major axis, 35.

ALPHABETICAL INDEX.

Eclipses (ἔκλειψις, a disappearance), 237-255; explanation of, 238; of the Moon, 241, 242; of the Sun, 243-245; annular, explained, 244; recurrence of, 247; phenomena attending a total eclipse of the Sun, 248; number of, 253; memorable, 254; effects on the ignorant, 255.

Ecliptic (so called because, when either Sun or Moon is *eclipsed*, it is in this circle), the great circle of the heavens, along which the Sun performs his annual journey, 358; plane of the, 111, 145; secular variation of the obliquity of the ecliptic, 458.

Egress, the passing of one body off the disk of another; *e. g.*, one of the satellites off Jupiter, or Venus off the Sun.

Egyptians, their early progress in Astronomy, 20.

Elements, chemical, present in the Sun, 126; in fixed stars, 83; in the Earth's crust, 212; in meteorites, 321.

Elements of an orbit, quantities the determination of which enables us to know the form and position of the orbit of a comet or planet, and to predict the positions of the body, *see* Appendix, Tables II.-V.

Ellipse, defined, 35; how it may be drawn, 35; major and minor axis, defined, 35; different forms of, 174; one of the conic sections, 513.

Elongation, the angular distance of a planet from the Sun, 372; of Mercury and Venus, 372.

Emersion, the reappearance of a body after it has been eclipsed or occulted by another; *e. g.*, the emersion of Jupiter's satellites from behind Jupiter, or of a star from behind the Moon.

Encke's Comet, 298, 299, 301, 304.
Envelopes of Comets, 301.
Ephemeris (ἐπί, for, and ἡμέρα, a day), a statement of the positions of the heavenly bodies for every day or hour, prepared some time beforehand, 463.

Epoch, some common period for which the positions of the heavenly bodies are calculated, 460.

Equation of the centre, the difference between the true and the mean anomaly of a planet or comet; of the equinoxes, the difference between the mean and apparent equinox; of time, the difference between true solar and mean solar time, 394.

Equator, of a sphere, 32; terrestrial, 162; celestial, 327.
Equinoctial Line, *see* Equator.
Equatorial, telescope, 436; method of using, 443; horizontal parallax, 455.

Equinoxes (*æquus*, equal, and *nox*, night), the points of intersection of the ecliptic and equator, 183; the Earth as seen from the Sun at the, 189; precession of the, 457; how produced, 528-530.

Eratosthenes, measured the Earth's circumference, 21.
Establishment of the port, 539.
Evection (*evehers*, to carry away). One of the moon's inequalities which increases or diminishes her mean longitude to the extent of 1° 20'.
Evening Star, 263.
Eye-pieces of telescopes, 431; transit eye-piece, 446.

Faculæ (Lat. *torches*), the brightest parts of the solar photosphere, 119.
Falling Bodies, velocity of, 499.
Field of view, the portion of the heavens visible in a telescope.
Fixed stars, *see* Stars.
Flora, the asteroid nearest the Sun, 283.
Focus (Lat. *hearth*), the point at which converging rays meet, 418.
Foci of an ellipse, 35.
Forces, parallelogram of, 497.
Fossils (Lat. *fossilis*, dug), 195.
Foucault, proves the Earth's rotation, 163, 164; determines the velocity of light, 410.
Fraunhofer's Lines, 479, 483.

Galaxy (γάλακτος, of milk), the Greek name for the Milky Way, or Via Lactea, 46.
Galileo, first used the telescope, 23, 427; construction of his telescope, 431.
Geocentric (γῆ, the Earth, and κέντρον, a centre), as viewed from the

ALPHABETICAL INDEX.

centre of the Earth; latitude and longitude, 355.
Gibbous (Lat. *gibbus*, hunched) Moon, 236.
Globes, use of the, 340; celestial, 61; rectifying the globe, 340, 347; globe, celestial, explains Sun's daily motion, 359, 360.
Gnomon (γνώμων, an index), a sundial, 385-387.
Granules on the solar surface, 120.
Gravitation, Universal, 502; the Moon's path, 506; Kepler's laws, 507; results of, 525; perturbations, 525; precession, 528; nutation, 532; tides, 534, 536.
Gravity (*gravis*, heavy), 498; measure of, on the Earth, 499; on the Sun and planets, 524; centre of, 506, 517.
Great Bear, the constellation, 53, 342.
Greeks, their additions to astronomical knowledge, 21.
Gregorian calendar, 403.
Gyroscope (γῦρος, a circle, and σκοπέω, I see), 166.

Halley's Comet, 301.
Harvest Moon, 363.
Head of comets, 300.
Heavens, how to observe the, 347-349.
Heliacal rising or setting of a star is when it just becomes visible in evening, or invisible in morning, twilight.
Heliocentric (ἥλιος, the Sun, and κέντρον, a centre), as seen from, or referred to, the centre of the Sun; latitude and longitude, 355.
Heliometer (ἥλιος, and μέτρον, a measure), a telescope with a divided object-glass, designed to measure small angular distances with great accuracy; so called because first used to measure the Sun.
Hemispheres (ἡμι, half, and σφαῖρα, a sphere), half the surface of the celestial sphere.
Herschel, Sir W., discovered the inner satellites of Saturn, 282; discovered Uranus, 140; principle of his reflector, 434.
Hindoos, their ideas of an eclipse, 255.
Hipparchus, catalogued the stars, 21.

Horizon (ὁρίζω, I bound), true or rational, 330; sensible, 161, 330.
Horizontal Parallax, 455.
Hour-angle, the angular distance of a heavenly body from the meridian.
Hour-circle, the circle attached to the equatorial telescope, by which right ascensions are indicated, 448.
Huggins, Mr., his spectroscopic observations, 487.
Huyghens, discovered the true nature of Saturn's rings, 282; his arrangement of lenses in eye-pieces, 431.
Hyperbola (ὑπερβολή), one of the conic sections, 513.

Immersion (*immergere*, to plunge into), the disappearance of one heavenly body behind another, or in the shadow of another.
Inclination of an orbit, the angle between the plane of the orbit and the plane of the ecliptic; of the Sun, 112; of the Earth, 180; of the axis of the planets, see Appendix, Table II.
Inequalities, Secular; perturbations of the celestial bodies so small that they become important only in a long period of time, 525.
Inferior Conjunction, 370.
Inferior Planets, 256.
Instruments, astronomical, 427-448.
Irradiation, 223.

Jets in comets, 301.
Jovicentric (*Jovis*, of Jupiter, and κέντρον, a centre), as seen from, or referred to, the centre of Jupiter.
Julian calendar, 403.
Jupiter, 273; distance from the Sun and period of revolution, 148; diameter, 150; volume, mass, and density, 157; polar compression, 273; seasons, 274; description of, 275, 276; belts of, 276; its rapid rotation, 276; probably surrounded by an immense atmosphere, 277; satellites, 278, 279.

Kepler's Laws, 507; second law, explained, 508, 514; third law, illustrated and proved, 510, 511.
Kirchhoff's investigations of spectra, 479, 481.

Latitude (*latitudo*, breadth), terrestrial, 168; how obtained, 465; celestial, 355; how obtained, 461; latitude of a place is equal to the altitude of the pole, 339; Geocentric, Heliocentric, Jovicentric, Saturnicentric, latitude as reckoned from the centre of the Earth, Sun, Jupiter, and Saturn.

Lens, what it is, 416; its action on a ray of light 416; kinds of lenses, 417; refraction by convex, 418; refraction by concave, 423; axis of a, 420; achromatic, 424; chromatic and spherical aberration of, 420.

Le Verrier, discovered Neptune, 141.

Librations of the Moon, 220.

Light, what it is, 408; velocity of, 15, 409; aberration of, 410; refraction and reflection, 411-413; dispersion, 414.

Limb, the edge of the disk of the Moon, Sun, or a planet.

Line, defined, 24; straight and curved, defined, 25; parallel lines, defined, 25; line of apsides, 407; of nodes, the imaginary line between the ascending and descending node of an orbit, 221.

Longitude (*longitudo*, length), terrestrial, 169; how determined at sea, 191, 468; in fixed observatories, 467; celestial, 355; how determined, 461; mean, the angular distance from the first point of Aries of a planet or comet supposed to move with a mean rate of motion; Geocentric, Heliocentric, Jovicentric, Saturnicentric, longitude as reckoned from the centre of the Earth, the Sun, Jupiter, and Saturn.

Lunar Distances, used to determine terrestrial longitudes, 468.

Lunation (*lunatio*), the period of the Moon's journey round the Earth, 399.

Luni-solar precession, *see* Precession.

Magellanic clouds, 47.
Magnitudes of stars, 40, 41.
Major axis, *see* Axis.
Mars, 266; distance from the Sun and period of revolution, 148; diameter, 150; volume, mass, and density, 157; day and year, 266; description of, 267; how presented to the Earth in different parts of its orbit, 269, 379-381; its land, water, and clouds, 270; its ice and snow, 271; seasons, 272; how its distance from the Earth is determined, 471.

Mass, the quantity of matter a heavenly body contains; of Sun, 109; how determined, 522; of planets, 157; how determined, 523, 526.

Maximiliana, the asteroid farthest from the Sun, 293.

Mean distance of a planet, etc., is half the sum of the aphelion and perihelion distances. This is equal to the semi-axis major of an elliptic orbit, 148. Mean anomaly, *see* Anomaly; mean obliquity is the obliquity unaffected by nutation; mean time, 395; mean Sun, 393.

Mercury, 257; distance from Sun and period of revolution, 148; phases, 257; orbit and apparent diameter, 258; heat and light, 259; seasons, 259; day and year, 260; density and force of gravity on its surface, 260; polar compression, 260; mountains, 261; distance from the Sun, 148; diameter, 150; relative volume, mass, and density, 157; elongation, 372.

Meridian (*meridies*, midday), the great circle of the heavens passing through the zenith of any place and the poles of the celestial sphere, 167.

Metals and metalloids, list of, 212; present in the Sun and stars, 10.

Meteorites, 317; how divided, 317; remarkable meteoric falls, 319; chemical composition of, 320, 321; structure of, 322.

Meteors (μετέωρον), luminous, their position in the system, 138; divisions of, 307; numbers seen in a star-shower, 307; explanation of star-showers, 308-311; the November ring, 308; radiant-point, 311; cause of brilliancy, 313; shape of orbits, 312; weight of, 314; velocity of, 313; August and April showers, 315; detonating meteors, 316; sporadic, 318.

Metius, invented the telescope, 427.

Micrometer (μικρός, small, and μέτρον, measure), an instrument with fine movable wires attached to eyepieces, to measure small angular distances, 439.

ALPHABETICAL INDEX. 307

Microscopes (μικρός, small, and σκοπέω, I see), of the transit-circle, 446.
Milky Way, 46; opinions of the ancients respecting it, 46; stars increase in number as they approach it, 48; nebulæ do not, 101.
Minor, planets, 137; axis, *see* Axis.
Mira, "the marvellous," 74.
Month, the, 399; length of the lunar and other months, 400.
Moon, why its shape changes, 12; size, 217; distance, 218; orbit, 218, 222; period of revolution, 219; librations, 220; nodes, 221, 247; Moon's path concave with respect to the Sun, 222; earth-shine, 223; brightness of Moon, 224; apparent difference of size, 225; description of surface, 227-231; no water or atmosphere, 232; rotation, 234; day, 234; phases, 235, 236; eclipses, 241, 242; apparent motions, 362; harvest Moon, 363; action of gravity on path of, 506; influence of, in producing precession, 530; produces nutation, 532; influence of, in producing tides, 536-538; elements of, *see* Appendix, Table V.
Moons, *see* Satellites.
Morning Star, 263.
Motion, proper, of stars, 63; apparent, of planets, 364-374; direct and retrograde, 373; laws of, 496, 497; resultant, 497; curvilinear, how produced, 500.
Mountains, lunar, heights of, 229.

Nadir (*natara*, to correspond), the point beneath the feet, 327.
Neap tides, 535.
Nebulæ (*nebula*, a cloud), why so called, 7; are masses of gas, 13, 102, 488; classification of, 93; light of, 99; variability of, 100; distribution of, 101; physical distribution of, 102; spectrum analysis of, 486.
Nebular hypothesis, 103, 216.
Nebulous stars, 98.
Neptune, 289; distance from the Sun and period of revolution, 148; diameter, 150; volume, mass, and density, 157; discovery of, 141; light, heat, and density, 289.
Newton, discovered the law of gravitation, 23, 501.
Night, succession of day and, 182; how to find the length of, 301.

Nodes (*nodus*, a knot), the points at which a comet's or planet's orbit intersects the plane of the ecliptic: one is termed the ascending, the other the descending, node, 221. Longitude of the node, the angular distance of the node from the first point of Aries. Line of, 221.
Nubecula Major, 47.
Nubecula Minor, 47.
Nucleus (Lat. *kernel*), of a comet, 300, 301; of sun-spots, 116.
Nutation (*nutatio*, a nodding), an oscillatory movement of the Earth's axis, due to the Moon's attraction on the equatorial protuberance, 532.

Object-glass of telescopes, 428; illuminating power of, 429; accuracy required in constructing, 430; largest object-glass, 433.
Obliquity of the ecliptic, variation of, 458.
Occultation (*occultare*, to hide), the eclipsing of a star or planet by the Moon or a planet.
Opposition. A superior planet is in opposition when the Sun, Earth, and the planet, are in the same straight line, with the Earth in the middle, 370.
Optical double stars, 69.
Orbit (*orbis*, a circle), the path of a planet or comet round the Sun, or of a satellite round a primary; of the planets, 174; of the Earth, 176; of the Moon, 222; of Mercury, 258; of comets, 298; of meteors, 312; inclinations and nodes of the planetary orbits, 376.

Parabola, a section of a cone parallel to one of its sides, 513.
Parabolic orbits of comets, 298.
Parallactic inequality, an irregularity in the Moon's motion, arising from the difference of the Sun's attraction at aphelion and perihelion.
Parallax (παράλλαξις, alternation), 453; correction for, 454, 455; equatorial horizontal, 455; of the Moon, 470; of Mars, 471; of the Sun, old and new value, 472; of the stars, 473, 474.
Parallels of latitude, 167; of declination, 328.

ALPHABETICAL INDEX.

Penumbra (*pæne*, almost, and *umbra*, a shadow), the half-shadow which surrounds the deeper shadow in an eclipse, 240; of sun-spots, 116.

Perigee (περί, near, and γῆ, the Earth). (1) The point in the Moon's orbit nearest the Earth, 218. (2) The position in which the Sun or other body is nearest the Earth.

Perihelion (περί, near, and ἥλιος, the Sun), the point in an orbit nearest the Sun, 175; distance, the distance of a heavenly body from the Sun at its nearest approach; longitude of, one of the elements of an orbit, the angular distance of the perihelion point from the first point of Aries; passage, the time at which a heavenly body makes its nearest approach to the Sun; distance of comets, 293; of planets, *see* Appendix, Table II.; change of, 407.

Peri-Jove, Saturnium, etc., the nearest approach of a satellite to Jupiter, Saturn, etc.

Period (περί, round, and ὁδός, a path), or periodic time, the time of a planet's, comet's, or satellite's revolution; synodic, the time in which a planet returns to the same position with regard to the Sun and Earth, 375.

Perturbations (*perturbare*, to interfere with), the effects of the attractions of the planets and satellites upon each other, consisting of variations in their motions and orbits, 525, 526.

Phases (φάσις, an appearance), the various appearances presented by the illuminated portions of the Moon (235), and inferior planets (257, 262, 369) in different parts of their orbits.

Photography (φῶς, light, and γραφή, a painting), celestial, 495.

Photosphere (φωτός, of light, and σφαῖρα, a sphere), of the stars, 82, 83; of the Sun, 124.

Physical constitution, of the stars, 82-84; of the Sun, 126.

Plane, defined, 28; of the ecliptic, 111, 145.

Planet (πλανήτης, a wanderer), a cool body revolving round a central incandescent one. *Planets* change their positions with regard to the stars, 5; what they are, 11; names and symbols of, 138; explanation of symbols, 139; historical details, 140; a suspected planet, 143; travel round the Sun in elliptical orbits, 144; their orbits lie nearly in the plane of the ecliptic, 145; motions of, 147; distances from the Sun, 148; periods of revolution, 148; diameters of, 150; comparative size of, 151; mass, volume, and density, 154-157; minor, 290-296; apparent movements of, 361-374; varying distances from the Earth, 365-367; variations in size and brilliancy, 368; phases, 369; aspects, 370; inferior and superior, 256; conjunction and opposition, 370; elongations, 372; direct and retrograde motion, and stationary points, 373, 374; synodic periods, 375; inclinations and nodes of orbits, 376; apparent paths among the stars, 379; elements of, *see* Appendix, Table II.

Planetary nebulæ, 97.

Pleiades, a star-group, 86.

Pointers, the, 342.

Polar, diameter of the Earth, 162, 171; circles, 170; compression of the Earth, explained, 202, 203; distance, 329; axis of equatorial, 436.

Polaris (*Lat.*), the pole-star, 341, 342; will not always mark the position of the north pole, 457.

Poles (πολέω, I turn), the extremities of the imaginary axis on which a celestial body rotates, 162; poles of the heavens, the extremities of the axis of the celestial sphere, 328; poles of the ecliptic, the extremities of the axis at right angles to the plane of the ecliptic, 360; of the Earth, 162; north celestial pole, 341; motion of, 457.

Pores, on the solar surface, 122.

Position-circle (of micrometers), 439.

Precession (*præcedere*, to precede) of the equinoxes, a slow retrograde motion of the equinoctial points upon the ecliptic, 356, 457; cause of, explained, 528-530.

Prime vertical, 333.

Prisms (πρίσμα), refract light, 413.

Prominences, red, of the Sun, 123, 251.

Proper motion, *see* Motion.

ALPHABETICAL INDEX. 309

Ptolemy, his theory of the solar system, 21; arranged the stars in 48 constellations, 54.
Punctulations, on the solar surface, 122.
Pythagoras, taught that the planets revolve round the Sun, 21; divined the truth respecting the Milky Way, 46.

Quadrant (*quadrans*, a fourth part), the fourth part of the circumference of a circle, or 90°; of altitude, a flexible strip of brass graduated into 90°, attached to the celestial globe, for determining celestial latitudes—declinations being determined by the brass meridian.
Quadrature. Two heavenly bodies are said to be in quadrature when there is a difference of longitude of 90° between them, 370.
Quarters of the Moon, 235.

Radiant-point of meteoric showers, 311.
Radiation, solar, 130.
Radius (*Lat.* a spoke of a wheel) vector, an imaginary line joining the Sun and a planet or comet in any point of its orbit, 508.
Rain, how caused, 211.
Red-flames, and prominences, 123, 251.
Reflecting telescope, 434; Earl of Rosse's, 435.
Reflection, 411.
Refracting telescope, 428-433.
Refraction (*refrangere*, to bend), atmospheric, 411, 412; of light by prisms, 413; correction for, 451.
Retrogradation, arc of, the arc apparently traversed by planets while their motion is retrograde, 373, 374.
Retrograde motion, 373, 374.
Revolution, the motion of one body round another; time of, the period in which a heavenly body returns to the same point of its orbit. The revolution may either be anomalistic if measured from the aphelion or perihelion point, sidereal with reference to a star, synodical with reference to a node, or tropical with reference to an equinox or tropic.

Right Ascension, 328.
Rilles, on the Moon, 231.
Rings of Saturn, 282-287; why sometimes invisible, 382.
Rocks, stratified, 194; list of, 194; igneous, 197.
Rotation, the motion of a body round a central axis: of Sun, 110; of Earth, 177; of Moon, 234; of Earth, possibly slackening, 511.
Rutherford, Mr., his lunar photographs, 495.

Saros, a term applied by the Chaldeans to the cycle of eclipses, 247.
Satellites (*satelles*, a companion), the smaller bodies revolving round planets and stars, 12, 146, 152; motions of, 147; of Jupiter, 278; of Saturn, 281; of Uranus, 288; satellite of Neptune, 289; elements of, *see* Appendix, Table III.
Saturn, 280; distance from the Sun and period of revolution, 148; diameter, 150; volume, mass, and density, 157; belts of, 281; satellites, 281; rings of, 282-287; dimensions of, 284; of what composed, 285; appearance of, from the body of the planet, 286; why sometimes invisible on the Earth, 382; atmosphere, 286; seasons, 286; solar eclipses due to the rings, 287; how presented to the Earth in different parts of its orbit, 382.
Schreibersite, a mineral found in meteorites, 320.
Scintillation (*scintilla*, a spark), the "twinkling" of the stars.
Seasons of the Earth, 186-189; of Mercury, 252; of Venus, 264; of Mars, 272; of Jupiter, 274.
Secular (*seculum*, an age), inequalities, 525; acceleration of the Moon's mean motion, 541.
Selenography (σελήνη, the Moon, and γράφω, I write), a description of the Moon.
Semi-diurnal arc, *see* Arc.
Sextant, an instrument consisting of the sixth part of a circle, finely graduated, with which the angular distances of celestial bodies are measured, 440.
Shooting-stars, *see* Meteors.

310   ALPHABETICAL INDEX.

Sidereal (*sidus*, a star), relating to the stars; clock, 397; day, 396; time, 397.
Signs of the zodiac, 356.
Sirius, its comparative brightness, 41.
Snow on Mars, 271.
Solar spectrum, 477.
Solar system, 137-158.
Solid, defined, 36.
Solstices, or solstitial points (*sol*, the Sun, and *stare*, to stand still), the points in the Sun's path at which the extreme north and south declinations are reached, and at which the motion is apparently arrested before its direction is changed, 183; the Earth as seen from the Sun at the, 188.
Solstitial colure, *see* Colure.
Spectroscope (*spectrum*, and σκοπέω, I see), 492; experiments with, 480; the Kew spectroscope, 493; direct vision, 494.
Spectrum, 414; the solar, 477; gradual formation of, 478; dark lines and bright lines, 479, 480; spectrum analysis, general laws of, 482; importance of, 485; spectra of the stars, 484; of nebulæ, 486; of Moon and planets, 489; solar, stellar, and nebular spectra, illustrated, *see* Frontispiece.
Sphere (σφαίρα), defined, 37; celestial, the sphere of stars which apparently encloses the Earth, 1, 326; of observation, 390.
Spheroid (σφαίρα, a sphere, and είδος, form), the solid formed by the rotation of an ellipse on one of its axes; it is oblate if it rotates on the minor axis, and prolate if it rotates on the major axis, 39.
Spring tides, 535.
Star-showers, *see* Meteors.
Stars, why invisible in daytime, 3; why they appear at rest, 8; why they shine, 10; distance of nearest, 15; their distance generally, 42, 44; magnitudes of, 40, 41; telescopic, 40; comparative brightness of, 41; distances of, 43, 44; diameters of, 45; distribution of, 48; divided into constellations, 53-57; names of, 58, 60; the twenty brightest, 62; proper motion of, 63; apparent motion of, 65, 334-337; double and multiple, 66-70; variable,

71-77; new or temporary, 76; the Sun a variable star, 125; colored, 78-80; colored double, 79; physical constitution of, 82-84; groups and clusters of, 86-88; nebulous, 98; apparent movements of, 334; apparent daily movements, 335-337; apparent yearly movements, 345, 346; visible in different latitudes, 338, 339; pole-star, 341, 342; those seen at midnight are opposite to the Sun, 346; how to identify the, 347; constellations visible throughout the year, 349-351; circumpolar, 338; sidereal day, 353; catalogues, 449; spectra of, 484; how the elements in the stars are determined, 485; parallax of the stars, 473, 474.
Stationary-points, points in a planet's orbit at which it appears to have no motion among the stars, 373, 374.
Stones, meteoric, 317.
Styles, old and new, 404; of sun-dials, 387.
Sun, is a star, 9; why it shines, 10; its relative brilliancy, 41, 106; approaching the constellation Hercules, 64; its disk, 105; distance, 107; diameter, 108; volume and mass, 109; rotation, 110; inclination of axis, 112; sun-spots, proper motion of, 113; description of, 115-118; size of, 125; period of, 125; telescopic appearance of, 114-123; photosphere, 116, 124; atmosphere, 124, 126; faculæ, 119; corrugations, or granules, 120; willow-leaves, 121; pores, or punctulations, 122; red-flames, 123, 251; the Sun, a variable star, 125; elements in the photosphere, 126; how determined by spectrum analysis, 483; benign influences of, 127; light, 128; heat, 129; chemical force, 131; habitability 134; future of, 135; eclipses of, 243-245; their phenomena, 248-252; solar heat, how accounted for by some, 313; apparent motions, 352; solar day, 353; motion in the ecliptic, 358; rising, setting, and apparent daily path, 358; daily motion, explained with the celestial globe, 360; mean Sun, 390-393; irregularities of the Sun's apparent daily motion, 391; distance, how determined, 472; parallax of, old and new

ALPHABETICAL INDEX. 311

value, 472; elements of, see Appendix, Table IV.
Sun-dial, the, 385-387.
Superior Conjunction, 370.
Superior Planets, 256.
Surface, defined, 28 plane, convex, and concave, defined, 28.
Symbols (σύμβολον), signs used as abbreviations, see Appendix, Table I.
Synodic period, 375.
Syzygies (σύν, with, and ζυγόν, a yoke), the points in the Moon's orbit at which it is in a line with the Earth and Sun, or when it is in conjunction or opposition.

Tails of comets, 300, 301.
Tangent, defined, 31.
Telescope (τῆλε, afar, and σκοπέω, I see), history, 427; construction, 428; illuminating or space-penetrating power, 99, 429; magnifying power, 430; eye-pieces, 428, 431; object-glass, 428-430; tube, 432; powers of, 433, 435; largest refractor, 433; reflecting, 434; largest reflector, 435; various mountings, 436; equatorial, 436; determination of positions with, 448; altazimuth, 442; transit-circle, 443-447; transit-instrument, 464.
Temperature, of the Sun, 129; of the Earth's crust, 188, 199.
Temporary Stars, 76.
Terminator, 228.
Thales, taught that the world was round, 21; predicted a memorable eclipse, 254.
Theophrastus, his opinion respecting the Milky Way, 46.
Tides (Saxon tidan, to happen), 534; spring and neap, 535; how produced, 536-539; velocity and height of tidal wave, 540; effect of tidal action on Earth's rotation, 541.
Time, as measured by the Sun, differs in different longitudes, 190; how to convert difference of time to difference of longitude, 191; how measured, 383; the mean Sun, motion of, 393; equation of time, 394; sidereal, 397; week, 398; month, 399; year, 401; bissextile, 403; Julian and Gregorian calendar, 403, 404; how determined, 464; table of, see Appendix, Table VI.

Trade-winds, 207-209.
Transit (trans, across, and ire, to go), the passage (1) of a heavenly body across the meridian of a place (in the case of circumpolar stars there is an upper and a lower transit, the latter sometimes called the transit sub polo); (2) of one heavenly body across the disk of another, e. g., the transit of Venus across the Sun, 371; of a satellite of Jupiter across the planet's disk, 278; methods of determining the time of, 447.
Transit-circle, when used and general description of, 443; determination of positions with, 444-446; determination of the time of transit, 447.
Transit-instrument, 464.
Triangle, defined, 30.
Tropics (τρέπω, I turn) of Cancer and Capricorn, the circles of declination which mark the most northerly and southerly points in the ecliptic, in which the Sun occupies the signs named, 170.
Tycho Brahe, added two constellations, 54; his clock, 389.

Ultra-zodiacal planets, a name sometimes given to the minor planets, because their orbits exceed the limits of the zodiac.
Umbra (Lat. a shadow), the darkest central portion of the shadow cast by a heavenly body, such as the Moon or Earth, 240; of sun-spots, 116.
Universe, our, one of many, 8; shape of, 49-51.
Uranus, 288; discovered by Sir Wm. Herschel, 140; distance from the Sun and period of revolution, 148; diameter, 150; volume, mass, and density, 157; satellites of, 288; specific gravity of, 288.

Vapor, aqueous, 215.
Variable, stars, 71-77; nebulæ, 100.
Variation of the Moon, one of the lunar inequalities.
Venus, 262; distance from the Sun and period of revolution, 148; diameter, 150; volume, mass, and density, 157; polar compression, 264; its

ALPHABETICAL INDEX.

phases, 262; seasons, 264; heat and light, 265; mountains, 265; a morning and evening star by turns, 263; path of, among the stars, 378; transits of, across the Sun's disk, 371, 472.

Verniers, 438.

Vertical (*vertex*, the top). A vertical line (331) is a line perpendicular to the surface of the Earth at any place, and is directed therefore to the zenith; a vertical circle is one that passes through the zenith and nadir of the celestial sphere; the prime vertical (333) is the vertical circle passing through the east and west points of the horizon.

Via Lactea (*Lat.*), *see* Milky Way.

Volcanoes, of the Moon, 227.

Volume (*volumen*, bulk), the cubical contents of a celestial body; of the Sun, 102; of the planets, 157.

Vulcan, a suspected planet, 143.

Walled Plains, on the Moon, 231.

Week, names of the days of, 398.

Weight, what it is, 498.

Willow-leaves in the penumbra of sun-spots, 121.

Winds, 206-210.

Year, the, 401; length of the sidereal and other years, 402; change in the length of solar, 400; length of the planets' years, *see* Appendix, Table II.

Zenith, the point of the celestial sphere overhead, 327; distance, 332.

Zodiac, the portion of the heavens extending 9° on either side of the ecliptic, in which the Sun and major planets appear to perform their annual revolutions, 356.

Zodiacal, light, 138; its shape, 307; how explained, 307; constellations, 55.

Zones, torrid, frigid, and temperate, 170.

THE END.

Astronomy.

Bowen's Astronomy by Observation.
By ELIZA A. BOWEN. . . . . $1.00.

An elementary text-book for high schools and academies, based on the most practical and interesting method of studying the subject—that of observation.

Gillet and Rolfe's First Book in Astronomy.
By JOSEPH A. GILLET and N. J. ROLFE. . . . . . $1.00.

This book, while intended for junior classes, is by no means primary or elementary. It is designed as a brief course, to serve as a foundation for more extended study.

Gillet and Rolfe's Astronomy.
By JOSEPH A. GILLET and N. J. ROLFE. . . . . . $1.40.

This book has been prepared by practical teachers, and contains nothing beyond the comprehension of the student of a high school or a seminary.

Kiddle's Short Course in Astronomy.
By HENRY KIDDLE, A. M. Fully illustrated. . . . $0.65.

This is a short course in Astronomy and the use of the globes. In mechanical execution it is unsurpassed.

Kiddle's New Elementary Astronomy.
By HENRY KIDDLE, A. M. . . . . $1.08.

A new manual of the elements of Astronomy, descriptive and mathematical, comprising the latest discoveries and theoretical views, with directions for the use of globes, and for studying the constellations.

Lockyer's Elementary Astronomy.
By J. N. LOCKYER, F. R. S. . . . . $1.22.

Accompanied with numerous illustrations, a colored representation of the solar, stellar, and nebular spectra, and Arago's celestial charts of the Northern and Southern Hemispheres. Especially adapted to the wants of American schools.

Ray's New Elements of Astronomy.
Revised edition. By S. H. PEABODY. . . . . . . $1.20.

The elements of Astronomy with numerous engravings and star maps. The author has restricted himself to plain statements of the facts, principles, and processes of the science.

Steele's New Descriptive Astronomy.
By J. DORMAN STEELE, Ph. D. . . . . $1.00.

This book is not written for the information of scientific men, but for the inspiration of youth. The author has sought to weave the story of those far-distant worlds into a form that may attract the attention and kindle the enthusiasm of the pupil.

Copies mailed, post-paid, on receipt of price. Full price-list sent on application.

AMERICAN BOOK COMPANY,
NEW YORK   ·:·   CINCINNATI   ·:·   CHICAGO.

PUBLICATIONS OF THE AMERICAN BOOK COMPANY.

Geology.

Andrews's Elementary Geology.
By E. B. ANDREWS, LL. D. . . . . $1.00.
This book is designed for students and readers of the Interior States, and therefore has its chief references to home geology. The scope is limited, to adapt it to beginners.

Dana's Geological Story Briefly Told.
By JAMES D. DANA, LL. D. . . . . $1.15.
With numerous illustrations. An introduction to geology for the general reader, and for beginners in the science. It contains a complete alphabetical index of subjects.

Dana's Manual of Geology.
By JAMES D. DANA, LL. D. . . . . $3.84.
This is a treatise on the principles of the science adapted to the wants of the American student, with special reference to American geological history. The illustrations are numerous, accurate, and well executed.

Dana's New Text-Book of Geology.
By JAMES D. DANA, LL. D. . . . . $2.00.
On the plan of the Manual, designed for schools and academies. The explanations are simple, and at the same time complete.

Le Conte's Compend of Geology.
By JOSEPH LE CONTE. . . . . $1.20.
A book designed to interest the pupil, and to convey real scientific knowledge. It cultivates the habit of observation by directing the attention of the pupil to scientific phenomena.

Nicholson's Text-Book of Geology.
By H. A. NICHOLSON. . . . . $1.05.
This presents the leading principles and facts of geological science within as brief a compass as is compatible with clearness and accuracy.

Steele's Fourteen Weeks in Geology.
By J. DORMAN STEELE, Ph. D. . . . . $1.00.
Designed to make science interesting by omitting those details which are valuable only to the scientific man, and by presenting only those points of general importance with which every well-informed person wishes to be acquainted.

Williams's Applied Geology.
By S. G. WILLIAMS. . . . . $1.20.
A treatise on the industrial relations of geological structure, and on the nature, occurrence, and uses of substances derived from geological sources. It gives a connected and systematic view of the applications of geology to the various uses of mankind.

Copies mailed, post-paid, on receipt of price. Full price-list sent on application.

AMERICAN BOOK COMPANY,
NEW YORK   ·:·   CINCINNATI   ·:·   CHICAGO.

PUBLICATIONS OF THE AMERICAN BOOK COMPANY.

Zoology and Natural History.

Cooper's Animal Life.
By SARAH COOPER. . . . . $1.25.

Animal life in the sea and on the land. A zoology for young people. Especial attention has been given to the structure of animals, and to the wonderful adaptation of this structure to their habits of life.

Holder's Elementary Zoology.
By C. F. HOLDER. . . . . $1.20.

A text-book designed to present in concise language the life-histories of the groups that constitute the animal kingdom, giving special prominence to distinctive characteristics and habits.

Hooker's Child's Book of Nature.
Part II. Animals. By WORTHINGTON HOOKER, M. D. $0.44.

While this work is well suited as a class-book for schools, its fresh and simple style can not fail to render it a great favorite for family reading.

Hooker's Natural History.
By WORTHINGTON HOOKER, M. D. . . . . $0.90.

For the use of schools and families. Illustrated by three hundred engravings. The book includes only that which every well-informed person ought to know, and excludes all which is of interest only to those who intend to be thorough zoologists.

Morse's First Book in Zoology.
By E. S. MORSE, Ph. D. . . . . $0.87.

Prepared for the use of pupils who wish to gain a general knowledge concerning the common animals of the country. The examples presented for study are such as are common and familiar to every school-boy.

Nicholson's Text-Book of Zoology.
By H. A. NICHOLSON, M. D. . . . . $1.38.

Revised edition. A work strictly elementary, designed for junior students. Illustrated with numerous engravings. It contains an Appendix, Glossary, and Index.

Steele's New Popular Zoology.
By J. DORMAN STEELE, Ph. D. . . . . $1.20.

This book proceeds, by natural development, from the lowest form of organism to man. A cut is given of every animal named, since a good picture of an object is worth more than pages of description.

Tenney's Elements of Zoology.
By SANBORN TENNEY, A. M. . . . . $1.60.

Illustrated by seven hundred and fifty wood engravings. It gives an outline of the animal kingdom, and presents the elementary facts and principles of zoology.

Tenney's Natural History of Animals.
By SANBORN TENNEY and ABBY A. TENNEY. . . . $1.20.

A brief account of the animal kingdom, for the use of parents and teachers. Illustrated by five hundred wood engravings, chiefly of North American animals.

Copies mailed, post-paid, on receipt of price. Complete price-list sent on application.

AMERICAN BOOK COMPANY,
NEW YORK ·:· CINCINNATI ·:· CHICAGO.

Physics and Natural Philosophy.

Arnott's Physics. Seventh edition. Edited by ALEXANDER BAIN, LL. D. . . . . $2.40.
An excellent and complete treatise on Natural Philosophy and Astronomy.

Cooley's New Text-Book of Physics. . . 90 cents.
An elementary course in Natural Philosophy designed for high schools and academies. It gives special prominence to the principle of energy.

Cooley's Easy Experiments. . . . . 52 cents.
A course of experiments in physical science for oral instruction in common schools. The experiments are such as intelligent boys and girls can make with little assistance.

Cooley's Elements of Natural Philosophy. 72 cents.
Revised edition. For common and high schools. It presents the most elementary facts of Natural Philosophy in an entertaining and instructive manner.

Everett's Outlines of Natural Philosophy. 84 cents.
Designed for schools and for general readers. It is at once easy enough for a class reading-book, and precise enough for a text-book

Gillet and Rolfe's First Book in Natural Philosophy. . . . . 60 cents.
For the use of schools and academies. It contains a brief, simple, and natural statement of the facts and principles of the science

Hooker's Child's Book of Nature. Part III. 44 cents.
A course of lessons on Air, Water, Heat, Light, etc., admirably illustrated.

Norton's Elements of Physics. . . . . . . 80 cents.
A systematic epitome of the science, designed as a text-book for academies and common schools.

Norton's Natural Philosophy. . . . . $1.10.
A selection of the facts and principles of Natural Philosophy, but adapted to the requirements of the pupil.

Peck's Ganot Revised. . . . . $1.20.
An introductory course in Natural Philosophy for the use of high schools and academies, edited from Ganot's Popular Physics by William G. Peck.

Quackenbos's Natural Philosophy. . . . . $1.22.
Designed to exhibit the application of scientific principles in every-day life.

Steele's Popular Physics. . . . . $1.00.
Designed both to interest and instruct the pupil. Written in simple language, and containing many practical illustrations.

Trowbridge's New Physics. . . . . $1.20.
A manual of experimental study for high schools and preparatory schools.

Wells's Natural Philosophy. . . . . $1.15.
This embodies the latest and best results of scientific discovery and research.

Copies mailed, post-paid, on receipt of price. Complete price-list sent on application.

AMERICAN BOOK COMPANY,
NEW YORK   ·:·   CINCINNATI   ·:·   CHICAGO.

PUBLICATIONS OF THE AMERICAN BOOK COMPANY.

Chemistry.

Brewster's First Book of Chemistry.
By MARY-SHAW BREWSTER. . . . . 66 cents.
A course of experiments of the most elementary character for the guidance of children in the simplest preliminary chemical operations. The simplest apparatus is employed.

Clarke's Elements of Chemistry.
By F. W. CLARKE. . . . . $1.20.
A class-book intended to serve not only as a complete course for pupils studying chemistry merely as part of a general education, but also as a scientific basis for subsequent higher study.

Cooley's New Elementary Chemistry for Beginners.
By LE ROY C. COOLEY. . . . . 72 cents.
This is emphatically a book of experimental chemistry. Facts and principles are derived from experiments, and are clearly stated in their order.

Cooley's New Text-Book of Chemistry.
By LE ROY C. COOLEY. . . . . 90 cents.
A text-book of chemistry for use in high schools and academies.

Eliot and Storer's Elementary Chemistry.
Abridged from Eliot and Storer's Manual, by WILLIAM RIPLEY NICHOLS, with the co-operation of the authors. . . . . . $1.08.
Adapted for use in high schools, normal schools, and colleges.

Steele's New Popular Chemistry.
By J. DORMAN STEELE, Ph.D. . . . . $1.00.
Devoted to principles and practical applications. Not a work of reference, but a pleasant study. Only the main facts and principles of the science are given.

Stoddard's Qualitative Analysis.
By JOHN T. STODDARD, Ph.D. . . . . 75 cents.
An outline of qualitative analysis for beginners. The student is expected to make the reactions and express them in written equations.

Stoddard's Lecture Notes on General Chemistry.
Part I. Non-Metals. . . . . $0.75.
Part II. Metals. . . . . 1.00.
Designed as a basis of notes to be taken on a first course of experimental lectures on general chemistry, to relieve the student from the most irksome part of his note-taking.

Youmans's Class-Book of Chemistry.
By EDWARD L. YOUMANS, M.D. Third edition. Revised and partly rewritten by WILLIAM J. YOUMANS, M.D. . . . . $1.22.
Designed as a popular introduction to the study of the science for schools, colleges, and general reading. With a colored frontispiece and 158 illustrations.

Copies mailed, post-paid, on receipt of price. Full price-list sent on application.

AMERICAN BOOK COMPANY,
NEW YORK　·:·　CINCINNATI　·:·　CHICAGO.

[*68]

General Science.

Doerner's Treasury of Knowledge.
By CELIA DOERNER. Part I. . . . . $0.50.
Part II. . . . . .65.

This book is designed to fill a gap in the ordinary course of instruction, and furnishes in a small compass much useful and important information. Since it combines entertainment with instruction, it will be found especially useful to parents as an addition to the child's home library.

Hooker's Child's Book of Nature. (COMPLETE.)
By WORTHINGTON HOOKER, M. D., ... $1.00.

Three parts in one: Part I. Plants; Part II. Animals; Part III. Air, Water, Heat, Light, etc. Designed to aid mothers and teachers in training children in the observation of Nature. It presents a general survey of the kingdom of Nature in a manner calculated to attract the attention of the child, and at the same time to furnish him with accurate and important scientific information.

Monteith's Easy Lessons in Popular Science.
By JAMES MONTEITH. . . . . $0.75.

This book combines the conversational, catechetical, blackboard, and object plans, with maps, illustrations, and lessons in drawing, spelling, and composition. The subjects are presented in a simple and effective style, such as would be adopted by a good teacher on an excursion with a class.

Monteith's Popular Science Reader.
By JAMES MONTEITH. . . . . $0.75.

This contains lessons and selections in Natural Philosophy, Botany, and Natural History, with blackboard, drawing, and written exercises. It is illustrated with many fine cuts, and brief notes at the foot of each page add greatly to its value.

Steele's Manual. (KEY TO FOURTEEN WEEKS' COURSE.)
By J. DORMAN STEELE, Ph. D. . . . . $1.00.

This is a manual of science for teachers, containing answers to the practical questions and problems in the author's scientific text-books. It also contains many valuable hints to teachers, minor tables, etc.

Wells's Science of Common Things.
By DAVID A. WELLS, A. M. . . . . $0.85.

This is a familiar explanation of the first principles of physical science for schools, families, and young students. Illustrated with numerous engravings. It is designed to furnish for the use of schools and young students an elementary text-book on the first principles of science.

Copies mailed, post-paid, on receipt of price. Full price-list sent on application.

AMERICAN BOOK COMPANY,

NEW YORK ·:· CINCINNATI ·:· CHICAGO.

Physical Geography.

In addition to the series of Political Geographies published by the American Book Company, their list includes the following standard and popular text-books on Physical Geography:

APPLETONS' PHYSICAL GEOGRAPHY.
Large 4to $1.60

Prepared by a corps of scientific experts with richly-illustrated engravings, diagrams, and maps in color, and including a separate chapter on the geological history and the physical features of the United States.

CORNELL'S PHYSICAL GEOGRAPHY.
Large 4to $1.12

Revised edition, with such alterations and additions as were found necessary to bring the work in all respects up to date.

ECLECTIC PHYSICAL GEOGRAPHY.
12mo $1.00

By RUSSELL HINMAN. A new work in a new and convenient form. All irrelevant matter is omitted and the pages devoted exclusively to Physical Geography clearly treated in the light of recent investigations. The numerous charts, cuts, and diagrams are drawn with accuracy, fully illustrating the text.

GUYOT'S PHYSICAL GEOGRAPHY.
Large 4to $1.60

By ARNOLD GUYOT. Revised, with new plates and newly-engraved maps. A standard work by one of the ablest of modern geographers. All parts of the subject are presented in their true relations and in their proper subordination.

MONTEITH'S NEW PHYSICAL GEOGRAPHY.
4to $1.00

A new and comprehensive work, embracing the results of recent research in this field, including Physiography, Hydrography, Meteorology, Terrestial Magnetism, and Vulcanology. The topical arrangement of subjects adapts the work for use in grammar grades as well as for high and normal schools.

Any of the above books will be mailed, postpaid, on receipt of price. Full price-list of books on all subjects for all grades will be sent on application.

AMERICAN BOOK COMPANY,
NEW YORK   ∴   CINCINNATI   ∴   CHICAGO

GEOGRAPHY.

STANDARD TWO-BOOK SERIES

All of these geographies have been edited with great care to present the latest information regarding the geography of the world and the results of the most recent researches and discoveries. They also have special editions for some or all of the States.

APPLETONS' STANDARD GEOGRAPHIES.

Appletons' Elementary Geography	55 cents
Appletons' Higher Geography	$1.25

The elementary book is objective in method. In the advanced, special prominence is given to industrial, commercial, and practical features.

BARNES'S NEW GEOGRAPHIES. By JAMES MONTEITH.

Barnes's Elementary Geography	55 cents
Barnes's Complete Geography	$1.25

A special feature of these books is the plan of teaching by comparison, or association of ideas. The advanced book includes Physical, Descriptive, Commercial, and Industrial Geography.

CORNELL'S GEOGRAPHIES. Revised Series.

Cornell's Primary Geography (New Edition)	42 cents
Cornell's Intermediate Geography (New Edition)	86 cents

These popular books have been entirely revised, yet the distinctive features of the series remain unchanged.

ECLECTIC GEOGRAPHIES. New Two-Book Series.

Eclectic Elementary Geography	55 cents
Eclectic Complete Geography	$1.20

The text of the Eclectic Geographies is comprehensive and simply worded. The maps show physical features of the earth's surface fully and accurately.

HARPER'S GEOGRAPHIES.

Harper's Introductory Geography	48 cents
Harper's School Geography	$1.08

The introductory book is written in pleasing narrative style. The advanced book presents physical and political elements in proper order of sequence, showing relations of industries and wealth to physical characteristics.

NILES'S GEOGRAPHIES.

Niles's Elementary Geography	44 cents
Niles's Advanced Geography	$1.00

The elementary book consists of interesting reading lessons, cultivating observation and exciting the imagination. The advanced book includes Mathematical, Physical, and Political Geography.

SWINTON'S GEOGRAPHIES.

Swinton's Introductory Geography	55 cents
In Readings and Recitations.	
Swinton's Grammar-School Geography	$1.25
Physical, Political, and Commercial.	

The text of these books is carefully graded so that the Introductory connects with the Grammar School without the need of any intermediate manual.

Any of the above books will be sent, postpaid, to any address on receipt of price. Correspondence in reference to the introduction of these books is cordially invited. Special allowances made when books are exchanged.

AMERICAN BOOK COMPANY

NEW YORK ∴ CINCINNATI ∴ CHICAGO

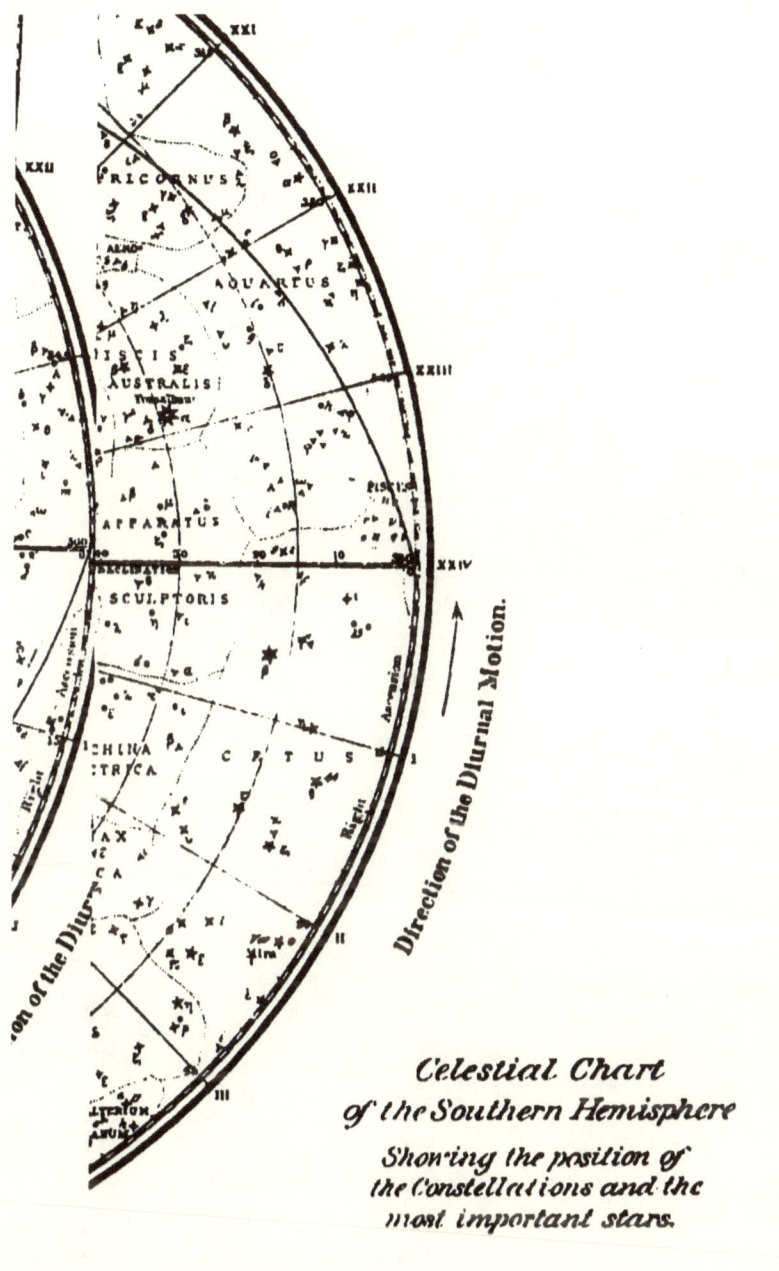

Celestial Chart of the Southern Hemisphere

Showing the position of the Constellations and the most important stars.

www.ingramcontent.com/pod-product-compliance
Lightning Source LLC
Chambersburg PA
CBHW030014240426
43672CB00007B/947